Lecture Notes in Geosystems Mathematics and Computing

Series editors
W. Freeden, Kaiserslautern
Z. Nashed, Orlando
O. Scherzer, Vienna

 Birkhäuser

More information about this series at http://www.springer.com/series/15481

S.L. Gavrilyuk • N.I. Makarenko •
S.V. Sukhinin

Waves in Continuous Media

S.L. Gavrilyuk
Aix-Marseille University
Marseille, France

N.I. Makarenko
Russian Academy of Sciences
Lavrentyev Institute of Hydrodynamics
Novosibirsk State University
Novosibirsk, Russia

S.V. Sukhinin
Russian Academy of Sciences
Lavrentyev Institute of Hydrodynamics
Novosibirsk State University
Novosibirsk, Russia

Translated by Tamara Rozhkovskaya
Sobolev Institute of Mathematics
Siberian Branch of the Russian Academy
   of Sciences
Novosibirsk, Russia

Lecture Notes in Geosystems Mathematics and Computing
ISBN 978-3-319-49276-6   ISBN 978-3-319-49277-3   (eBook)
DOI 10.1007/978-3-319-49277-3

Library of Congress Control Number: 2017930812

Printed on acid-free paper

This book is published under the trade name Birkhäuser, www.birkhauser-science.com
The registered company is Springer International Publishing AG
The registered company address is: Gewerbestrasse 11, 6330 Cham, Switzerland

Preface

Wave phenomena occur everywhere in nature and therefore are studied in many areas of science for a long time.

The mathematical wave theory emerged as an independent discipline in the mid-1970s due to numerous applications in natural science and engineering stimulating the further development of mathematical methods.

The lecture course *Waves in Continuous Media* is one of the disciplines on continuum mechanics and mathematical modeling included into the education program at the Department of Mechanics and Mathematics, Novosibirsk State University. This course was first given by Professor L. V. Ovsyannikov,[1] a distinguished scientist who obtained a number of fundamental results in the field of wave hydrodynamics. Based on Ovsyannikov's principles of selecting the material, the authors developed new variants of the course adapted to groups of master's students specialized in applied mathematics, mechanics, and geophysics.

The textbook contains a rich collection of exercises and problems which have been carefully selected and tested at practical works and seminars of courses given by the authors at Novosibirsk State University (Russia) and Aix-Marseille University (France) for many years. Most of the problems and exercises are supplied with answers and hints. Solutions of some typical problems are explained in detail, and some theoretical background material is included in order to make the book self-contained and give students the necessary tools for self-education. More than 200 problems formulated in the book allowed us to propose to each master's student an individual semester mini-project consisting in solving up to six problems. Most of them are solved by applying the theoretical approaches from the course, but the other ones demand a deeper understanding of the methods discussed in the course. During the semester, the students have also been working in research laboratories, so a set of problems specific to the research activity of the students was usually proposed.

---

[1]Ovsyannikov, L. V.: Wave Motions of Continuous Media. Novosibirsk State University, Novosibirsk (1985) [in Russian].

The textbook consists of three chapters. Chapters 1 and 2 present the basic notions and facts of the mathematical theory of waves illustrated by numerous examples and methods of solving typical problems. The reader learns how to recognize the hyperbolicity property; find characteristics, Riemann invariants, and conservation laws for quasilinear systems of equations; construct and analyze solutions with weak or strong discontinuities; and investigate equations with dispersion: analysis of dispersion relations, the study of large time asymptotic behavior of solutions, the construction of traveling wave solutions for models reducible to nonlinear evolution equations, etc. The majority of problems are formulated within the framework of wave models arising in gas dynamics, magnetohydrodynamics, elasticity and plasticity, linear and nonlinear acoustics, chemical adsorption, and other applications.

Chapter 3 deals with surface and internal waves in an incompressible fluid. The efficiency of mathematical methods is demonstrated on a hierarchy of approximate submodels generated from the Euler equations of homogeneous and inhomogeneous fluids. Some problems illustrate the influence of viscosity and vorticity on the wave processes.

The list of references consists mainly of monographs and textbooks recommended for further reading. Three of them are generic [1–3], while others [4–33] are more specific for each chapter. These have been selected to allow readers to understand better mathematical statements whose proofs were skipped, and find solutions of relatively hard exercises. A separate bibliography for each chapter is maintained. The reference list for Chap. 3 also contains five research articles on the theory of water waves [16, 17, 21, 26, 31] we explicitly refer to. The books for further reading are not cited in the text.

The authors thank their colleagues at the Chair of Hydrodynamics, Novosibirsk State University, for their help in the preparation of the manuscript.

The authors would like to express their special gratitude to Professor V. M. Teshukov, who recently passed away, for his numerous useful advices and discussions.

Marseille, France   S.L. Gavrilyuk
Novosibirsk, Russia   N.I. Makarenko
Novosibirsk, Russia   S.V. Sukhinin
October 2016

Contents

Chapter 1
Hyperbolic Waves

1.1 Hyperbolic Systems

We consider the quasilinear system of first order equations

$$\mathbf{u}_t + A(\mathbf{u}, x, t)\mathbf{u}_x + \mathbf{b}(\mathbf{u}, x, t) = 0, \tag{1.1}$$

where the $n \times n$-matrix A and vector $\mathbf{b}$ depend on x, t and $\mathbf{u} = (u_1, \ldots, u_n)^T$. A direction $dx/dt = c$ is called *characteristic* if there exists a linear combination of equations of the form (1.1) such that each unknown function u_i is differentiable along this direction. The quantity c in the definition of a characteristic direction is an eigenvalue of the matrix A, i.e.,

$$\det(A - cI) = 0. \tag{1.2}$$

For any eigenvalue c and the corresponding left eigenvector $\mathbf{l} = (l_1, \ldots, l_n)$ of the matrix A (i.e., $\mathbf{l}A = c\mathbf{l}$) the system (1.1) implies the following condition on the *characteristic* (the curve corresponding to a characteristic direction):

$$\mathbf{l} \cdot (d_t\mathbf{u} + \mathbf{b}) = 0, \tag{1.3}$$

where $d_t = \partial_t + c\partial_x$ is the operator of differentiation along the characteristic.

The system (1.1) is *hyperbolic* if all eigenvalues c_i of the matrix A are real (in this case, they can be ordered: $c_1 \leqslant c_2 \leqslant \ldots \leqslant c_n$) and there exist n linearly independent real left eigenvectors of the matrix A.

Figure 1.1 shows the location of characteristics emanating from a given point M in the (x, t)-plane. A hyperbolic system of equations is equivalent to a system of n relations on characteristics. A system is hyperbolic if and only if the normal Jordan form of the matrix A is diagonal. We indicate the sufficient hyperbolicity

© Springer International Publishing AG 2017
S.L. Gavrilyuk et al., *Waves in Continuous Media*, Lecture Notes in Geosystems Mathematics and Computing, DOI 10.1007/978-3-319-49277-3_1

Fig. 1.1 Location of
characteristics emanating
from a given point M in the
(x, t)-plane

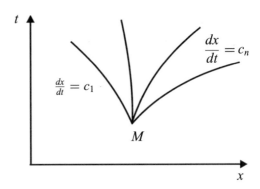

conditions:

(a) the matrix A is symmetric,
(b) all roots of Eq. (1.2) are real and distinct.

In case (b), where the matrix A has no multiple eigenvalues, the system (1.1) is called *strictly hyperbolic*.

Example 1.1 The process of chemical adsorption used for separating substances in a liquid or gas mixture by the chromatography method is described by the equations

$$\partial_t \left(\mathbf{u} + \mathbf{f}(\mathbf{u}) \right) + v \partial_x \mathbf{u} = 0, \tag{1.4}$$

where $\mathbf{u} = (u_1, \dots, u_n)^T$ are the concentrations of the separated substances passing through the adsorption column, $\mathbf{f}(\mathbf{u}) = (f_1(u), \dots, f_n(u))^T$ are the concentrations of the substances adsorbed by the adsorbent, and $v = \text{const} > 0$ is the mixture velocity. Let the vector-valued function $\mathbf{f}(\mathbf{u})$, called the *adsorption isotherm*, be such that all eigenvalues of the Jacobi matrix

$$\mathbf{f}'(\mathbf{u}) = \frac{\partial(f_1, \dots, f_n)}{\partial(u_1, \dots, u_n)}$$

are real, positive, and distinct:

$$0 < \lambda_1 < \dots < \lambda_n.$$

Then Eqs. (1.4) can be transformed to the form (1.1) with the matrix $A(\mathbf{u}) = v(I + \mathbf{f}'(\mathbf{u}))^{-1}$ and vector $\mathbf{b} = 0$. Since

$$A - cI = ((v - c)I - c\mathbf{f}'(\mathbf{u}))(I + \mathbf{f}'(\mathbf{u}))^{-1},$$

the eigenvalues of the matrix A are connected with the eigenvalues of the matrix $\mathbf{f}'(\mathbf{u})$ by the identity

$$c_j = \frac{v}{1 + \lambda_j} \quad (j = 1, 2, \ldots, n).$$

Consequently, the system (1.4) is strictly hyperbolic; moreover, all its characteristic velocities are positive and do not exceed the mixture velocity. The noncoincidence of the velocities $c_i \neq c_j \ (i \neq j)$ is the basis of the chromatography method.

If there exist scalar functions $r(\mathbf{u})$ and $\mu(\mathbf{u}, x, t)$ such that

$$\frac{\partial r}{\partial u_i} = \mu l_i \quad (i = 1, \ldots, n),$$

then the relation (1.3) is equivalent to the equation

$$d_t r(\mathbf{u}) = -\mu \mathbf{l} \cdot \mathbf{b},$$

where $r(\mathbf{u})$ is called a *Riemann invariant*. The motivation of this definition becomes clear in the case $\mathbf{l} \cdot \mathbf{b} = 0$, where the Riemann invariant r is constant along characteristics. Riemann invariants always exist for the system (1.1) consisting of one or two equations and for the system (1.1) with constant matrix A of arbitrary order n (in the second case, $r = \mathbf{l} \cdot \mathbf{u}$). In the general case $n \geq 3$, Riemann invariants do not necessarily exist. In the case $n = 3$, the identity $\mathbf{l} \cdot \text{curl}\,\mathbf{l} = 0$ is necessary and sufficient for the existence of a Riemann invariant for the characteristic $dx/dt = c$ corresponding to a simple eigenvalue c of the matrix A with eigenvector $\mathbf{l}(\mathbf{u}) = (l_1, l_2, l_3)$.

Problem 1.1 Find characteristics and Riemann invariants for the system describing shallow water flows over the flat bottom

$$\begin{aligned} h_t + (uh)_x &= 0, \\ u_t + uu_x + gh_x &= 0, \end{aligned} \tag{1.5}$$

where $h(x, t)$ is the layer depth, $u(x, t)$ is the horizontal fluid velocity, and g is the acceleration of gravity.

Solution We compose the matrix of coefficients of the original system of equations

$$A - cI = \begin{pmatrix} u - c & h \\ g & u - c \end{pmatrix}.$$

Then we find the characteristic velocities $c^\pm = u \pm \sqrt{gh}$. The system is hyperbolic in the domain $h > 0$. For the characteristic $dx/dt = c^+$ the left eigenvector, defined up to an arbitrary scalar factor, has the form $\mathbf{l} = (\sqrt{g}, \sqrt{h})$. Consequently, to find the Riemann invariant $r(h, u)$, we should find a solution to the system of equations

$$\frac{\partial r}{\partial h} = \mu \sqrt{g},$$

$$\frac{\partial r}{\partial u} = \mu \sqrt{h},$$

where $\mu(h, u)$ is an unknown integrating factor. Excluding this factor, we obtain the linear first order partial differential equation for r

$$\frac{\partial r}{\partial u} - \sqrt{\frac{h}{g}} \frac{\partial r}{\partial h} = 0.$$

From the equation of characteristics $du = -\sqrt{g/h}\,dh$ we find the first integral $r = u + 2\sqrt{gh}$. Since there is a certain functional arbitrariness in the definition of a Riemann invariant, the obtained first integral can be taken for the sought invariant. The characteristic $dx/dt = c^-$ is studied in a similar way.   □

Answer:

$$\frac{dx}{dt} = u + \sqrt{gh}: \quad u + 2\sqrt{gh} = \text{const},$$

$$\frac{dx}{dt} = u - \sqrt{gh}: \quad u - 2\sqrt{gh} = \text{const}.$$

1.2   Propagation of Weak Discontinuities

The Cauchy problem for the system (1.1) is to find a solution for $t > t_0$ provided that $u_i(x, t_0) = u_{i0}(x)$ is given at $t = t_0$.

Theorem 1.1 (uniqueness) *Assume that the system* (1.1) *is hyperbolic and the coefficient matrix A and vector* **b** *are continuously differentiable. Let a continuously differentiable solution* **u**(x, t) *be defined in the characteristic triangle* $X_1 M X_n$ *(cf. Fig.* 1.2*). If* $\bar{\mathbf{u}}$ *is another continuously differentiable solution to the system* (1.1) *in* $X_1 M X_n$ *and* $\bar{\mathbf{u}} = \mathbf{u}$ *on the segment* $X_n X_1$, *then* $\bar{\mathbf{u}} = \mathbf{u}$ *in the entire characteristic triangle* $X_1 M X_n$.

This theorem implies the existence of wave fronts defined by the characteristics $X_1 M$ and $X_n M$ and defining the *domain of determinacy* of the solution to the Cauchy problem by the initial data solely at the *domain of dependence* $X_n X_1$ of the point M.

Let a domain D be divided by a smooth curve $\Gamma : x = \chi(t)$ into two subdomains D_- and D_+ (cf. Fig. 1.3).

Fig. 1.2 The solution to the Cauchy problem for the system (1.1) is uniquely determined by the initial data on the interval $X_n X_1$ inside the curvilinear characteristic triangle $X_1 M X_n$

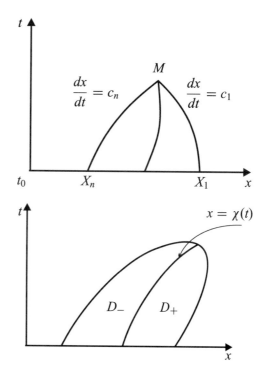

Fig. 1.3 The weak discontinuities always propagate along the characteristic curves

We assume that the solution to the hyperbolic system is continuous in the closed domain $\overline{D}$ and continuously differentiable in the closures $\overline{D}_-$ and $\overline{D}_+$. Moreover, the derivative $\partial_x \mathbf{u} = \mathbf{v}$ of the solution on Γ can have a discontinuity of the first kind with jump $[\mathbf{v}] = \mathbf{v}_+ - \mathbf{v}_-$. By the continuity of the solution $\mathbf{u}$, the jump of its tangent derivative $d_t \mathbf{u} = \partial_t \mathbf{u} + \chi'(t)\partial_x \mathbf{u}$ on Γ vanishes, which implies the following expressions for the jumps of derivatives:

$$[\partial_x \mathbf{u}] = [\mathbf{v}],$$
$$[\partial_t \mathbf{u}] = -\chi'[\mathbf{v}].$$

Consequently, from the system (1.1) we have

$$(A - \chi' I)[\mathbf{v}] = 0.$$

Thus, the derivative of the solution can be discontinuous only on the characteristic; moreover, the jump is a right eigenvector of the matrix A. In the case of a simple eigenvalue of the matrix A, the amplitude of the weak discontinuity is characterized by a scalar σ such that $[\mathbf{v}] = \sigma \mathbf{r}$, where $\mathbf{r} = (r_1, \ldots, r_n)^T$ is the corresponding right eigenvector. The quantity σ satisfies the ordinary differential equation along the characteristic

$$(\mathbf{l} \cdot \mathbf{r}) \frac{d\sigma}{dt} + P\sigma + Q\sigma^2 = 0, \tag{1.6}$$

where $\mathbf{l}$ is the corresponding left eigenvector, whereas P and Q are known functions. In particular,

$$Q = \sum_{i,j,k=1}^{n} l_i \left(\frac{\partial a_{ij}}{\partial u_k} \right) r_k r_j,$$

where a_{ij} are components of the matrix A. Without loss of generality we can assume that $(\mathbf{l} \cdot \mathbf{r}) = 1$. The relation (1.6) is the Riccati equation. It is called the *transport equation for the amplitude of the weak discontinuity*.

Problem 1.2 Consider the system of equations describing the isentropic motion of a polytropic gas, written in terms of the Riemann invariants r and l,

$$\begin{cases} r_t + (u+c)r_x = 0, \\ l_t + (u-c)l_x = 0, \end{cases} \quad r = u + \frac{2}{\gamma-1}c, \quad l = u - \frac{2}{\gamma-1}c,$$

with the initial conditions

$$u(x,0) = \begin{cases} 0, & x \geq a, \\ c_0(x-a)/(l_0 + a - x), & x < a, \end{cases} \quad c(x,0) = c_0,$$

where $a = \text{const}$, $c_0 = \text{const}$, and $l_0 = \text{const}$ ($c_0 > 0$, $l_0 > 0$). Compute the jump $[u_x]$ of the derivative on the characteristic $x = c_0 t + a$ at time t.

Solution By the uniqueness theorem for the Cauchy problem, $u(x,t) \equiv 0$ and $c(x,t) \equiv c_0$ for $x \geq c_0 t + a$. Further, for the Riemann invariant l along the characteristic $dx/dt = u + c$ we have

$$[l_t] + (u-c)[l_x] = 0,$$
$$[l_t] + (u+c)[l_x] = 0,$$

where first relation immediately follows from the equation of motion, whereas the second one is obtained from the continuity of the tangent derivative of l on the weak discontinuity line. Therefore, along the characteristic under consideration, we have $[l_t] = 0$, $[l_x] = 0$, and, as a consequence, $[u_x] = [r_x]/2$. Differentiating the first equation of the original system with respect to x, considering the jump, and taking into account the above properties $l_x(x-0,t) = l_x(x+0,t) = 0$ for $x = c_0 t + a$, we obtain the transport equation

$$\frac{d[r_x]}{dt} - \frac{\gamma + 1}{4}[r_x]^2 = 0$$

with the condition $[r_x] = -c_0/l_0$ at $t = 0$ which follows from the initial data. Integrating the equation, we obtain an expression for the jump $[u_x]$.   □

Answer:

$$[u_x] = -\frac{c_0}{2l_0 + \frac{\gamma+1}{2}c_0 t}.$$

1.3   Motion with Strong Discontinuities

The construction of solutions describing the shock wave propagation is based on the conservation laws

$$\partial_t \boldsymbol{\varphi}(x, t, \mathbf{u}) + \partial_x \boldsymbol{\psi}(x, t, \mathbf{u}) = \mathbf{f}(x, t, \mathbf{u}), \qquad (1.7)$$

where $\boldsymbol{\varphi} = (\varphi_1, \ldots, \varphi_n)^T$, $\boldsymbol{\psi} = (\psi_1, \ldots, \psi_n)^T$, and $\mathbf{f} = (f_1, \ldots, f_n)^T$ are n-dimensional vectors interpreted as densities, fluxes, and sources of the sought quantities. Not every system of differential equations (1.1) can be written in the conservative form (1.7). For models of continuous media this property automatically follows from the main equations written as a system of integral conservation laws on an arbitrary interval $[x_1, x_2]$:

$$\frac{d}{dt} \int_{x_1}^{x_2} \boldsymbol{\varphi}(x, t, \mathbf{u}) dx + \boldsymbol{\psi}(x, t, \mathbf{u}) \Big|_{x=x_1}^{x=x_2} = \int_{x_1}^{x_2} \mathbf{f}(x, t, \mathbf{u}) dx. \qquad (1.8)$$

Systems of the form (1.1) representable as systems of independent conservation laws with nonzero Jacobian $|\partial \boldsymbol{\varphi}/\partial \mathbf{u}| \neq 0$ are called *conservative*. In many cases, the number of conservation laws is larger than the number of unknowns. In such a situation, a right choice of a system of conservation laws for describing motions with strong discontinuities should be done with taking into account additional properties (stability of solutions, existence of a discontinuous solution as the limit of smooth solutions, physical interpretation of the limit of smooth solutions, physical interpretation of conservation laws, and so on).

Problem 1.3 Find all scalar conservation laws $\partial_t \varphi(u, v) + \partial_x \psi(u, v) = 0$, where the density φ is a polynomial in v of degree at most 2, for the system of nonlinear elasticity

$$u_t = v_x,$$

$$\rho_0 v_t = \sigma_x,$$

$$\sigma = \sigma(u),$$

$$\rho_0 = \text{const}.$$

Solution According to the definition of a conservation law, we have

$$\varphi_u u_t + \varphi_v v_t + \psi_u u_x + \psi_v v_x = 0.$$

Excluding the derivatives u_t and v_t by using the original system and then collecting and equating to zero the coefficients at u_x and v_x, we obtain the system of equations for φ and ψ

$$\rho_0 \psi_u + \sigma'(u)\varphi_v = 0,$$

$$\psi_v + \varphi_u = 0.$$

Applying the cross differentiation, we exclude the function ψ and obtain the linear second order partial differential equation for the density φ

$$\rho_0 \varphi_{uu} = \sigma'(u)\varphi_{vv}.$$

In this problem, it is required to find all solutions of the form

$$\varphi = \alpha(u)v^2 + \beta(u)v + \gamma(u)$$

to this equation. For the coefficients of the above polynomial we have the relations $\alpha'' = 0$, $\beta'' = 0$, and $\rho_0 \gamma'' = 2\sigma'\alpha$. Integrating these relations, we obtain the solution

$$\varphi = C_1 \varphi_1 + \ldots + C_5 \varphi_5,$$

where C_j are arbitrary real constants and φ_j are the basic densities specified below.   □

Answer:

$$\varphi_1 = u, \quad \varphi_2 = \rho_0 v, \quad \varphi_3 = uv,$$

$$\varphi_4 = \frac{1}{2}\rho_0 v^2 + \int_0^u \sigma(\xi)d\xi, \quad \varphi_5 = \frac{1}{2}\rho_0 uv^2 + \int_0^u (2\xi - u)\sigma(\xi)d\xi.$$

Fig. 1.4 The solutions can
have a discontinuity of the
first kind on the line $x = X(t)$

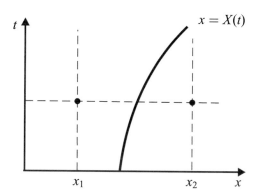

Assume that a solution **u** has a discontinuity of the first kind on the line $x = X(t)$ and is smooth on both sides of this line. We fix integration limits x_1 and x_2 in the integral conservation law (1.8) such that $x_1 < X(t) < x_2$ at a given time t (cf. Fig. 1.4). Dividing the integration interval by the point $x = X(t)$ into two parts, differentiating the obtained integrals with respect to the time, and passing to the limit as $x_1 \rightarrow x_2$, we obtain the relations on the strong discontinuity, called the *Rankine–Hugoniot conditions*,

$$D[\boldsymbol{\varphi}] = [\boldsymbol{\psi}], \tag{1.9}$$

where $D = dX(t)/dt$ is the velocity of the strong discontinuity. We fix a state on one side of the discontinuity given by the point $\mathbf{u} = \mathbf{u}_0$ in $\mathbb{R}^n$. Then, on the other side of the wave, the locus of admissible states is the curve in $\mathbb{R}^n$ given by (1.9). This curve, called the *shock adiabat*, can consist of several smooth branches passing through the center $\mathbf{u}_0$.

Example 1.2 Consider the system (1.7) of conservation laws in nonlinear elasticity with $\boldsymbol{\varphi} = (u, v)$, $\boldsymbol{\psi} = -(v, \sigma(u)/\rho_0)$, and $\mathbf{f} = 0$. We assume that the stress function $\sigma(u)$ is such that $\sigma'(u) > 0$, $\sigma''(u) < 0$, $\sigma(0) = 0$, and $\sigma'(0) = \lambda + 2\mu$ (here, $\lambda > 0$ and $\mu > 0$ are the Lamé constants). We write the relation (1.9) on the strong discontinuity connecting the states (u_0, v_0) and (u, v) and exclude the wave velocity D. As a result, we obtain the shock adiabat equation

$$\rho_0(v - v_0)^2 = \{\sigma(u) - \sigma(u_0)\}(u - u_0). \tag{1.10}$$

This curve in the (u, v)-plane (u is the strain and v is the velocity of the material) is the locus of states obtained after the shock wave passes through the state (u_0, v_0). The shock adiabat (1.10) consists of two branches describing the shock propagating to the left ($v < v_0$) and to the right ($v > v_0$). Respectively, for the wave velocity we obtain the expression

$$D = \pm \sqrt{\frac{1}{\rho_0} \frac{\sigma(u) - \sigma(u_0)}{u - u_0}}.$$

We consider an elastic semi-infinite rod $x \geqslant 0$ in the equilibrium state under loading $\sigma_0 = \sigma(u_0)$. When the loading is suddenly removed from the end $x = 0$, the rod comes back to the unstrained state $u = 0$, $\sigma = 0$ due to the unloading shock wave travelling to the right with the velocity $D = \sqrt{\sigma(u_0)/(\rho_0 u_0)}$. For small initial strains u_0 the shock wave velocity is approximately equal to the velocity $c_0 = \sqrt{(\lambda + 2\mu)/\rho_0}$ of the linear longitudinal elastic wave.

1.4   Kinematic Waves

By *kinematic waves* we mean the class of one-dimensional motions of a continuous medium with a given dependence $q = Q(\rho)$ of the flow on the density. The knowledge of such a dependence allows us to obtain a closed model of motion by using only the law of conservation of mass

$$\rho_t + q_x = 0. \tag{1.11}$$

For particular media the relation between the flow and density is usually found from experiments or by integrating other equations of a more general model. In the kinematic-wave approximation, the description of motion is reduced to finding solutions to the quasilinear first order partial differential equation

$$\rho_t + c(\rho)\rho_x = 0,$$

where $c(\rho) = Q'(\rho)$ is the characteristic velocity.

Example 1.3 Consider the model of kinematic waves in a one-way traffic flow. A twice continuously differentiable function Q on $0 \leqslant \rho \leqslant \rho_*$ is characterized by the properties

(a) $Q(\rho) > 0$ $(0 < \rho < \rho_*)$,

(b) $Q''(\rho) < 0,$ (1.12)

(c) $Q(0) = Q(\rho_*) = 0.$

A typical graph of $Q(\rho)$ is shown in Fig. 1.5.

In this model of a continuous medium, the velocity of "particles" (individual vehicles) is equal to $u = Q(\rho)/\rho$. The value $\rho_* > 0$ yields the limit vehicle density on the highway when the vehicles stand bumper to bumper, so that no motion is

Fig. 1.5 Flow-density relationship

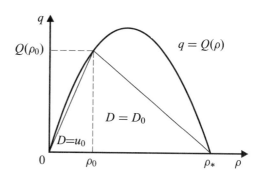

possible in view of the second identity in (1.12c). On the other hand, according to the first identity in (1.12c), $\lim_{\rho \to 0} Q(\rho)/\rho > 0$ exists and is equal to the velocity of the motion over the free highway. From (1.12b) it follows that $c'(\rho) < 0$, so that the characteristic velocity $c(\rho)$ is a monotonically decreasing function of the density ρ. Perturbations with abrupt fronts can propagate through moving vehicles when a sudden deceleration occurs somewhere in the traffic flow.

In the case of the conservation law (1.11), the relation on the strong discontinuity has the form

$$D[\rho] = [q]$$

or, in detail,

$$D = \frac{Q(\rho_2) - Q(\rho_1)}{\rho_2 - \rho_1}. \tag{1.13}$$

The Riemann problem for the kinematic wave equation is the Cauchy problem with piecewise constant initial data

$$\rho(x, 0) = \begin{cases} \rho_1, & x < 0, \\ \rho_2, & x > 0, \end{cases}$$

where $\rho_i = \text{const}$, $\rho_1 \neq \rho_2$. This problem has a solution in the class of self-similar motions $\rho = \rho(x/t)$ with strong and weak discontinuities. For $\rho_1 < \rho_2$ the solution is piecewise constant and has a strong discontinuity on the line $x = Dt$, where D is given by (1.13). According to this formula, we have

$$D = \frac{1}{\rho_2 - \rho_1} \int_{\rho_1}^{\rho_2} c(\rho)\,d\rho,$$

which implies $c_1 > D > c_2$, $c_i = Q'(\rho_i)$, provided that $c(\rho)$ is monotonically decreasing. The strong discontinuities satisfying the above condition are stable. If $\rho_1 > \rho_2$, then the velocities of shock wave propagation satisfy the opposite inequalities $c_1 < D < c_2$ which means that the strong discontinuities are unstable under small perturbations of the initial data. In this case, there exists a continuously stable solution representable as a centered wave with the following distribution of the characteristic velocity:

$$c(x,t) = \begin{cases} c_1, & x < c_1 t, \\ x/t, & c_1 t < x < c_2 t, \\ c_2, & x > c_2 t. \end{cases}$$

Thus, the choice of a stable solution to the Riemann problem for the kinematic wave equation is determined by the sign of the difference $\Delta\rho = \rho_1 - \rho_2$.

Problem 1.4 The traffic flow moves with velocity $u_0 > 0$ and density ρ_0 along a street where the traffic light, located at the point $x = 0$, turns red at time $t = 0$. In the kinematic-wave approximation, describe the initial stage of the traffic flow for $t > 0$ in a neighborhood of the traffic light.

Solution In the motion for $t > 0$ at the left from the traffic light, the following two states should be connected: the incoming traffic flow with velocity u_0 far from the traffic light and the convoy of immobile vehicles with density ρ_* in close vicinity of the traffic light. This situation is simulated by the problem with discontinuous initial-boundary conditions at the point $x = 0$, $t = 0$

$$\rho(x,0) = \rho_0 \ (x < 0), \quad \rho(0,t) = \rho_*.$$

Since $\rho_0 < \rho_*$, a shock wave appears in motion (the wave caused by a sudden halt of vehicles) which propagates in the direction opposite to the traffic flow direction. By formula (1.13) and the property (1.12c), the wave velocity is equal to $D_0 = -Q(\rho_0)/(\rho_* - \rho_0) < 0$. This quantity is the slope of the chord joining the points $(\rho_0, Q(\rho_0))$ and $(\rho_*, 0)$ (cf. Fig. 1.5). Further, on a road part at the right from the traffic light, there are no vehicles for $t > 0$. This state should be agreed with the outgoing traffic flow of vehicles passed by the traffic light before turning on the red light. This situation is described by the problem with discontinuous data $\rho(x,0) = \rho_0 \ (x > 0)$, $\rho(0,t) = 0$. Here, a shock wave appears that travels to the right with velocity $D = (Q(\rho_0) - Q(0))/\rho_0$. Since $Q(0) = 0$, the wave front corresponds to the location of the tail of the convoy of vehicles outgoing from the cross-road with constant velocity $u_0 = Q(\rho_0)/\rho_0$. The quantity u_0 is the slope of the chord joining the points $(0,0)$ and $(\rho_0, Q(\rho_0))$ in Fig. 1.5. The above-described stage of the traffic motion near the traffic light corresponds to a part of (x,t)–diagram in Fig. 1.6 observed on the time interval $0 < t < T_k$ of the red light. $\square$

Fig. 1.6 (x, t) - diagram
showing the traffic motion
near traffic lights

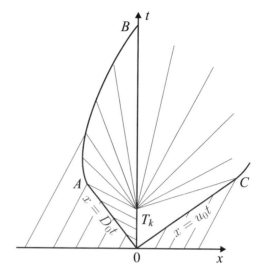

Answer:

$$u = u_0, \quad \rho = \rho_0 \ (x < D_0 t), \quad u = 0, \quad \rho = \rho_* \ (D_0 t < x < 0),$$
$$u = 0, \quad \rho = 0 \ (0 < x < u_0 t), \quad u = u_0, \quad \rho = \rho_0 \ (x > u_0 t).$$

1.5 Multi-dimensional Wave Fronts

Consider the linear system of equations

$$\partial_t \mathbf{u} + \sum_{i=1}^{3} A^i \partial_{x_i} \mathbf{u} + B \mathbf{u} = 0 \tag{1.14}$$

for an n-dimensional vector $\mathbf{u} = (u_1, \ldots, u_n)^T$, where $A^i(\mathbf{x}, t)$ and $B(\mathbf{x}, t)$ are given
$n \times n$-matrices depending on t and $\mathbf{x} = (x_1, x_2, x_3)^T$. We consider the hyperplane

$$\xi_1 x_1 + \xi_2 x_2 + \xi_3 x_3 + \tau t = \text{const}$$

in $\mathbb{R}^4$ with normal vector $\mathbf{v} = (\xi_1, \xi_2, \xi_3, \tau)^T$. A direction $\mathbf{v}$ is called *characteristic*
if

$$\det \left(\tau I + \sum_{i=1}^{3} \xi_i A^i \right) = 0. \tag{1.15}$$

The system (1.14) is said to be *hyperbolic* at a point $(\mathbf{x}, t)$ if for any $\boldsymbol{\xi} = (\xi_1, \xi_2, \xi_3)^T$ Eq. (1.15) has n real roots $\tau_k = H_k(\boldsymbol{\xi}; \mathbf{x}, t)$ $(k = 1, \ldots, n)$ and the characteristic matrix

$$A(\boldsymbol{v}) = \tau I + \sum_{i=1}^{3} \xi_i A^i$$

has n linearly independent vectors $\mathbf{l} \in \mathbb{R}^n$ such that $\mathbf{l} \cdot A(\boldsymbol{v}) = 0$.

A smooth hypersurface in $\mathbb{R}^4$ such that the tangent hyperplane at each point has the characteristic direction is called a *characteristic* of the system (1.14). Let the characteristic corresponding to the root $\tau = H(\boldsymbol{\xi}; \mathbf{x}, t)$ be given by the equation

$$\varphi(x_1, x_2, x_3, t) = 0.$$

Then its normal $\boldsymbol{v} = (\varphi_{x_1}, \varphi_{x_2}, \varphi_{x_3}, \varphi_t)^T$ has the characteristic direction. Therefore, φ satisfies the nonlinear first order partial differential equation, called the *Hamilton–Jacobi equation*,

$$\varphi_t = H(\nabla_{\mathbf{x}}\varphi; \mathbf{x}, t). \tag{1.16}$$

Characteristics of this equation are called *bicharacteristics* (rays in wave theory) of the system (1.14). The differential equations for bicharacteristics have the form of the Hamilton system of equations

$$\frac{dx_i}{dt} = -\frac{\partial H}{\partial p_i},$$

$$\frac{dp_i}{dt} = \frac{\partial H}{\partial x_i}$$

where $i = 1, 2, 3$, $p_i = \partial\varphi/\partial x_i$. For the linear hyperbolic system (1.14) with constant matrices A^i the function H is independent of $\mathbf{x}$ and t, and, consequently, $dp_i/dt = 0$, so that the right-hand sides $H_{p_i}(p_1, p_2, p_3)$ of the differential equations for x_i are constant along the rays. Consequently, the rays themselves are rectilinear:

$$x_i = x_i^{(0)} - H_{p_i}^{(0)} t.$$

Problem 1.5 The front of a two-dimensional acoustic wave propagating through a gas at rest with the sound speed c_0 has at $t = 0$ the shape of a parabola $y = x^2$. Find the time T at which the sound reaches the observer at the point $A(3, 0)$.

Solution The acoustic equations for a gas at rest are written in the form

$$\rho_0 u_t + p_x = 0,$$
$$\rho_0 v_t + p_y = 0,$$
$$p_t + \rho_0 c_0^2 (u_x + v_y) = 0,$$

where u and v are components of the velocity vector and p is the pressure perturbation. We write this system in the matrix form (1.14) and compose the characteristic determinant with the normal vector $\nu = (\xi, \eta, \tau)^T$:

$$\det (\tau I + \xi A^x + \eta A^y) = \begin{vmatrix} \tau & 0 & \xi/\rho_0 \\ 0 & \tau & \eta/\rho_0 \\ \rho_0 c_0^2 \xi & \rho_0 c_0^2 \eta & \tau \end{vmatrix} = \tau(\tau^2 - c_0^2(\xi^2 + \eta^2)) = 0.$$

Hence for the sound characteristics we have $H(p, q) = \pm c_0 \sqrt{p^2 + q^2}$, where $p = \varphi_x$, $q = \varphi_y$ ($\varphi(x, y, t) = 0$ is the location of the front at time t). Since the function H is independent of x and y, the bicharacteristics are rectilinear. Integrating, we obtain the equations of acoustic rays in the form

$$x = x_0 \pm \frac{p_0 c_0 t}{\sqrt{p_0^2 + q_0^2}},$$

$$y = y_0 \pm \frac{q_0 c_0 t}{\sqrt{p_0^2 + q_0^2}}.$$

Consequently, at each time moment, the bicharacteristics are directed along the normal to the front; moreover, perturbations propagate along the rays with constant velocity c_0. This means that the first perturbation reaching the observer is that outgoing from the point $B(x, x^2)$ lying on the initial location of the front and is the nearest to A. The minimum of distance $|AB|$ is attained at a point B with the x-coordinate satisfying the extremum condition $2x^3 + x - 3 = 0$. Such a point is unique and is $B(1, 1)$. The corresponding distance is equal to $\sqrt{5}$.   □

Answer: $T = \sqrt{5}/c_0$.

If the Hamiltonian H is independent of t, then the equation of characteristic surfaces can be looked for in the form $t = \psi(\mathbf{x})$. By Eq. (1.16) with $\varphi = \psi(\mathbf{x}) - t$, the function ψ satisfies the equation

$$H(\nabla_\mathbf{x} \psi, \mathbf{x}) = -1,$$

called the *eikonal equation*. In this case, the system for bicharacteristics is the autonomous system

$$\frac{dx_i}{ds} = \frac{\partial H}{\partial p_i},$$

$$\frac{dp_i}{ds} = -\frac{\partial H}{\partial x_i},$$

where $i = 1, 2, 3$, $p_i = \partial \psi / \partial x_i$, and s is the parameter along the bicharacteristics.

1.6 Symmetrization of Hyperbolic Systems of Conservation Laws

Consider the system of quasilinear equations

$$A(\mathbf{u})\partial_t \mathbf{u} + \sum_{i=1}^{3} B^i(\mathbf{u})\partial_{x_i} \mathbf{u} = 0 \tag{1.17}$$

for a vector-valued function $\mathbf{u} = (u_1, \ldots, u_n)^T$ depending on t and $\mathbf{x} = (x_1, x_2, x_3)^T$. It is assumed that the square $n \times n$-matrices $A(\mathbf{u})$ and $B^i(\mathbf{u})$ are symmetric and the matrix $A = (a_{ij})_{i,j=1}^{n}$ is positive definite:

$$\sum_{i,j=1}^{n} a_{ij} p_i p_j > 0$$

for any vector $\mathbf{p} = (p_1, \ldots, p_n)^T$, $\mathbf{p} \neq 0$. The system (1.17) with such matrices is called a *symmetric t-hyperbolic in the sense of Friedrichs system* (or simply *t-hyperbolic Friedrichs system*). In particular, the system (1.1) with two independent variables t, x and a symmetric matrix A is a symmetric t-hyperbolic Friedrichs system.

The t-hyperbolicity and conservation properties play an important role in the analysis of qualitative properties of solutions and numerical solving of such systems. Therefore, it is useful to know how to reduce a given hyperbolic system of quasilinear equations to the symmetric form (1.17). Such a reduction is possible in the case described by the following assertion.

Theorem 1.2 (Godunov, Friedrichs, and Lax) *Let the system of conservation laws*

$$\partial_t \mathbf{u} + \sum_{i=1}^{3} \partial_{x_i} \psi^i(\mathbf{u}) = 0 \tag{1.18}$$

admits the additional conservation law

$$\partial_t e(\mathbf{u}) + \sum_{i=1}^{3} \partial_{x_i} f^i(\mathbf{u}) = 0, \qquad (1.19)$$

where the function $e(\mathbf{u})$ is convex in the variables $\mathbf{u} = (u_1,\ldots,u_n)^T$, i.e., the Hessian matrix $e''(\mathbf{u}) = (\partial_{u_i}\partial_{u_j} e(\mathbf{u}))_{i,j=1}^{n}$ is positive definite. Then the system (1.18) is reduced to the form (1.17).

Proof The simultaneous validity of Eqs. (1.18) and (1.19) implies the following compatibility condition:

$$\nabla_{\mathbf{u}} f^i(\mathbf{u}) = \nabla_{\mathbf{u}} e(\mathbf{u})(\psi^i(\mathbf{u}))' \quad (i = 1,\ldots,n) \qquad (1.20)$$

for scalar-valued functions f^i and vector-valued functions ψ^i. We introduce the Legendre transform $e^*(\mathbf{v})$ of a function $e(\mathbf{u})$ by the formula

$$e^*(\mathbf{v}) = \mathbf{v} \cdot \mathbf{u} - e(\mathbf{u}),$$

where the vector $\mathbf{u}$ is implicitly defined by the equation $\mathbf{v} = \nabla_{\mathbf{u}} e(\mathbf{u})$. It is possible to invert the dependence of $\mathbf{v}$ on $\mathbf{u}$ since the Jacobian

$$\left| \frac{\partial(v_1,\ldots,v_n)}{\partial(u_1,\ldots,u_n)} \right| = \det e''(\mathbf{u})$$

differs from zero for convex functions $e(\mathbf{u})$. It is obvious that $\mathbf{u} = \nabla_{\mathbf{v}} e^*(\mathbf{v})$. Furthermore, $e^*(\mathbf{v})$ is a convex function of $\mathbf{v}$. We introduce the functions

$$f^{i*}(\mathbf{v}) = \mathbf{v} \cdot \psi^i(\mathbf{u}) - f^i(\mathbf{u}) \quad (i = 1,\ldots,n).$$

By (1.20), we have $\psi^i(\mathbf{u}) = \nabla_{\mathbf{v}} f^i(\mathbf{v})$. Consequently, the system (1.18) can be written as

$$\partial_t \nabla_{\mathbf{v}} e^*(\mathbf{v}) + \sum_{i=1}^{3} \partial_{x_i} \nabla_{\mathbf{v}} f^{i*}(\mathbf{v}) = 0,$$

i.e., in the form (1.17) with the Hessian matrices $A(\mathbf{v}) = (\partial_{v_k}\partial_{v_j} e^*(\mathbf{v}))_{k,j=1}^{n}$ and $B^i(\mathbf{v}) = (\partial_{v_k}\partial_{v_j} f^{i*}(\mathbf{v}))_{k,j=1}^{n}$. $\qquad\square$

The proof of this theorem contains a constructive symmetrization method for systems of conservation laws.

Example 1.4 Consider the system of equations with two independent variables t and x

$$h_t + (uh)_x = 0,$$

$$(hu)_t + \left(hu^2 + \frac{1}{2}gh^2 \right)_x = 0. \tag{1.21}$$

This system is the conservative form of the shallow water equations (1.5) for the laws of conservation of mass and momentum. For the additional conservation law (1.19) we take the law of conservation of energy

$$\partial_t e(h, u) + \partial_x f(h, u) = 0$$

with the functions

$$e = \frac{1}{2}u^2h + \frac{1}{2}gh^2,$$

$$f = \frac{1}{2}u^3h + guh^2.$$

We first write Eqs. (1.21) in the original form (1.18) with the vector of conservative quantities $\mathbf{u} = (u_1, u_2)$, where $u_1 = h$ and $u_2 = uh$. In this notation, the functions e, f and the components of vector $\boldsymbol{\psi} = (\psi_1, \psi_2)$ are expressed as

$$e(\mathbf{u}) = \frac{u_2^2}{2u_1} + \frac{1}{2}gu_1^2,$$

$$f(\mathbf{u}) = \frac{u_2^3}{2u_1^2} + gu_1^2u_2,$$

$$\psi_1(\mathbf{u}) = u_2, \quad \psi_2(\mathbf{u}) = \frac{u_2^2}{u_1} + \frac{1}{2}gu_1^2.$$

Since the function $e(\mathbf{u})$ is convex in $\mathbf{u}$, we have

$$v_1 = e_{u_1} = gu_1 - \frac{u_2^2}{u_1^2},$$

$$v_2 = e_{u_2} = \frac{u_2}{u_1}.$$

Inverting the dependence $\mathbf{v} = \nabla_{\mathbf{u}}e(\mathbf{u})$, we find

$$u_1 = \frac{1}{g}\left(v_1 + \frac{1}{2}v_2^2 \right),$$

$$u_2 = \frac{1}{g}v_2\left(v_1 + \frac{1}{2}v_2^2 \right).$$

Consequently, the functions $e^* = \mathbf{v} \cdot \mathbf{u} - e$ and $f^* = \mathbf{v} \cdot \boldsymbol{\psi} - f$ take the form

$$e^*(\mathbf{v}) = \frac{1}{2g}\left(v_1 + \frac{1}{2}v_2^2\right)^2,$$

$$f^*(\mathbf{v}) = \frac{1}{2g}v_2\left(v_1 + \frac{1}{2}v_2^2\right)^2.$$

Computing the Hessian matrices for these functions, we find the symmetric form of the system (1.21)

$$A(\mathbf{v})\mathbf{v}_t + B(\mathbf{v})\mathbf{v}_x = 0$$

with matrices

$$A = \begin{pmatrix} 1 & v_2 \\ v_2 & v_1 + \frac{3}{2}v_2^2 \end{pmatrix}, \quad B = \begin{pmatrix} v_2 & v_1 + \frac{3}{2}v_2^2 \\ v_1 + \frac{3}{2}v_2^2 & 3v_1 + \frac{5}{2}v_2^2 \end{pmatrix}.$$

There are other symmetric forms of hyperbolic systems of equations. For example, for systems with two independent variables one can obtain such a form by transformation of the original equations to equations written in Riemann invariants (if they exist). Thus, for the shallow water equations (1.5) we have

$$r_t + (u + \sqrt{gh})r_x = 0,$$

$$l_t + (u - \sqrt{gh})l_x = 0,$$

where $r = u + 2\sqrt{gh}$ and $l = u - 2\sqrt{gh}$. However, this approach is less general since it requires the existence of Riemann invariants.

1.7 Problems

1. Find a velocity field $u(x, t)$ for the one-dimensional motion of a continuous medium, where all particles are moving by inertia provided that, at the initial time $t = 0$, the medium occupies the half-space $x \geqslant 0$ and the velocity distribution has the form

a. $u(x, 0) = x^2$,
b. $u(x, 0) = \sqrt{x}$.

Answer:

a. $u(x, t) = \dfrac{2x^2}{2xt + 1 + \sqrt{4xt + 1}}$,

b. $u(x, t) = \dfrac{\sqrt{4x + t^2} - t}{2}$.

2. Integrate the equation of characteristics for solutions to the Cauchy problem

$$u_t + uu_x + u = 0,$$
$$u(x, 0) = ax + b,$$

where a and b are constants.
Answer: $x = x_0 + (ax_0 + b)(1 - e^{-t})$.

3. Construct a solution to the Cauchy problem

$$u_t + c(u)u_x = 0,$$
$$u(x, 0) = c^{-1}(ax + b),$$

where a and b are constants and $c(u)$ is a monotone smooth function. For what values of a and b does the gradient catastrophe occur?
Answer:

$$u(x, t) = c^{-1}\left(\frac{ax + b}{1 + at}\right).$$

4. Find characteristics and Riemann invariants of the system of equations

$$u_t + 2\cos v u_x + \sin u v_x = 0,$$
$$v_t + \cos v u_x + (\sin u + \cos v)v_x = 0.$$

Answer:

$$\frac{dx}{dt} = \cos v : \quad u - v = \text{const}$$

$$\frac{dx}{dt} = \sin u + 2\cos v : \quad \frac{1 + \sin v}{\cos v}\tan\frac{u}{2} = \text{const}.$$

5. Show that the Jacobian $|\partial(r_1, \ldots, r_n)/\partial(u_1, \ldots, u_n)|$ of the transformations $\mathbf{r} = \mathbf{r}(\mathbf{u})$ reducing the hyperbolic system of equations $\mathbf{u}_t + A(\mathbf{u})\mathbf{u}_x = 0$ to a system in Riemann invariants

$$\mathbf{r}_t + C(\mathbf{r})\mathbf{r}_x = 0, \tag{1.22}$$

where $C(\mathbf{r}) = \text{diag}(c_1, \ldots, c_n)$ is a diagonal matrix, is different from zero.

6. Consider the strictly hyperbolic system (1.22) reduced to Riemann invariants (i.e., $c_i \neq c_j$ for $i \neq j$). Prove that for a nondegenerate simple wave $\mathbf{r} = \mathbf{r}(\alpha(x, t))$, $\mathbf{r}'(\alpha) \neq 0$, all except one invariants r_i are identically constant, whereas the level lines of the simple wave $\alpha(x, t) = \text{const}$ form a family

of rectilinear characteristics corresponding to a Riemann invariant that is not identically constant.

7. For the hyperbolic system of equations

$$u_t + c(u, v)u_x = 0,$$
$$v_t + c(u, v)v_x = 0$$

with multiple characteristic $dx/dt = c(u, v)$ prove that $u_x > v_x > 0$ along this characteristic provided that these inequalities are valid at $t = 0$.

8. The one-dimensional motion of a barotropic continuous medium is governed by the system of equations

$$\rho_t + u\rho_x + \rho u_x = 0,$$
$$u_t + uu_x + \frac{c^2(\rho)}{\rho}\rho_x = 0,$$

where $c(\rho)$ is a smooth function such that $c'(\rho) > 0$ and $c(0) = 0$. Find a function $c = c(\rho)$ for which all characteristics of this system for any motion are straight lines.

Answer: $c = A\rho$, where $A = \text{const} > 0$.

9. Consider the system of equations of one-dimensional motions of a mixture of two barotropic fluids

$$\rho_t + u\rho_x + \rho u_x = 0,$$
$$u_t + uu_x + \frac{1}{\rho}p_x = 0, \qquad (1.23)$$
$$y_{1t} + uy_{1x} = 0,$$

where u is the mixture velocity, $\rho = \alpha_1\rho_1 + \alpha_2\rho_2$ is the mixture density (here, ρ_i are the densities and α_i are the volume fractions of the fluids; $0 \leqslant \alpha_i \leqslant 1, \alpha_1 + \alpha_2 = 1$), $p = p_1(\rho_1) = p_2(\rho_2)$ is the pressure ($dp_1/d\rho_1 > 0, dp_2/d\rho_2 > 0$ for $\rho_1 > 0, \rho_2 > 0$), and y_1 is the mass fraction of the first fluid ($y_i = \alpha_i\rho_i/\rho$). Show that the system of equations (1.23) for ρ, u, y_1 with an implicitly given pressure $p = p(\rho, y_1)$ is hyperbolic and its characteristics have the form $dx/dt = u$, $dx/dt = u \pm c$, where the sound speed c in the mixture (the Wood speed) is given by

$$\frac{1}{\rho c^2} = \frac{\alpha_1}{\rho_1 c_1^2} + \frac{\alpha_2}{\rho_2 c_2^2},$$
$$c_i^2 = \frac{dp_i}{d\rho_i} \quad (i = 1, 2).$$

Hint. Use the relation $\tau = y_1\tau_1 + y_2\tau_2$ for the specific volumes $\tau = 1/\rho$, $\tau_1 = 1/\rho_1$, and $\tau_2 = 1/\rho_2$.

10. Under the conditions of the previous problem, we consider an air-water mixture with parameters $\rho_1 = 1\,\text{kg/m}^3$, $\rho_2 = 1000\,\text{kg/m}^3$, $c_1 = 340\,\text{m/sec}$, and $c_2 = 1500\,\text{m/sec}$. Show that the sound speed $c = c(\alpha_1)$ in the mixture has a unique minimum $c_{\min}$ over the interval $0 \leqslant \alpha_1 \leqslant 1$. Find this minimum. *Answer:* $c_{\min} \approx 21.5\,\text{m/sec}$.

11. Show that any hyperbolic system (1.1) with two independent variables t and x can be reduced to the following form by the left multiplication by a suitable matrix:

$$B\mathbf{u}_t + C\mathbf{u}_x + D\mathbf{b} = 0,$$

where B, C, and D are symmetric matrices depending on $\mathbf{u}$, x, and t; moreover, the matrix B is positive definite (recall that such a system is called a symmetric t-hyperbolic Friedrichs system).

12. Let $\mathbf{u} = (u_1, \ldots, u_n)^T \in \mathbb{R}^n$, and let $e : \mathbb{R}^n \to \mathbb{R}$ be a given smooth mapping such that the Hessian matrix $e''(\mathbf{u}) = \|\partial_{u_i}\partial_{u_j}e''(\mathbf{u})\|_{i,j=1}^n$ is positive definite, i.e., $e(\mathbf{u})$ is a convex function. We consider the Legendre transform e^* of a function e defined by

$$e^*(\mathbf{v}) = \mathbf{v} \cdot \mathbf{u} - e(\mathbf{u}),$$

where $\mathbf{v} = \nabla_{\mathbf{u}}e(\mathbf{u})$ and $\mathbf{u} = \mathbf{u}(\mathbf{v})$ is the preimage of an element $\mathbf{v}$ under the action of the locally invertible mapping $\mathbf{v} = \nabla_{\mathbf{u}}e(\mathbf{u})$. Show that $e^*(\mathbf{v})$ is a convex function of $\mathbf{v}$. Show that the Legendre transform is an involution, i.e., double application of this transform yields the same function e. *Hint.* Show that $\mathbf{u} = \nabla_{\mathbf{v}}e^*(\mathbf{v})$ and $(e^*(\mathbf{v}))'' = (e''(\mathbf{u}))^{-1}$.

13. Consider the equations of one-dimensional motion of an ideal gas with zero pressure

$$\rho_t + (\rho u)_x = 0,$$

$$u_t + uu_x = 0.$$

(This approximation appears in astrophysics.) Show that this system is not hyperbolic. Find all conservation laws $\partial_t P(\rho, u) + \partial_x Q(\rho, u) = 0$ admitted by this system. Find out whether there are laws with a convex function P among these conservation laws. *Answer:* $P(\rho, u) = a(u)\rho + b(u)$, where a and b are arbitrary smooth functions; P is not convex.

14. Consider the system of conservation laws of gas dynamics

$$\rho_t + (\rho u)_x = 0,$$
$$(\rho u)_t + (p + \rho u^2)_x = 0,$$
$$(\rho s)_t + (\rho u s)_x = 0,$$
$$p = p(\rho, s),$$

where ρ is the density, u is the velocity, p is the pressure, and s is the entropy. The thermodynamical state of the medium is characterized by the internal gas energy $\varepsilon(\rho, s)$ and temperature $T(\rho, s)$ connected by the Gibbs identity

$$Tds = d\varepsilon + pd\left(\frac{1}{\rho}\right).$$

Show that the system of gas dynamics admits the additional conservation law

$$e_t + f_x = 0,$$

where

$$e = \rho\left(\varepsilon + \frac{1}{2}u^2\right),$$

$$f = \rho u\left(\varepsilon + \frac{1}{2}u^2\right) + pu.$$

15. Under the conditions of the previous problem, compute the Hessian matrix $e''(\mathbf{u})$ of the function $e = \rho(\varepsilon + u^2/2)$ in the variables $\mathbf{u} = (u_1, u_2, u_3)^T$, where $u_1 = \rho$, $u_2 = \rho u$, and $u_3 = \rho s$.
Answer:

$$e''(\mathbf{u}) = \frac{1}{\rho}\begin{pmatrix} u^2 + K & -u\,\rho\varepsilon_{\rho s} - s\varepsilon_{ss} \\ -u & 1 & 0 \\ \rho\varepsilon_{\rho s} - s\varepsilon_{ss} & 0 & \varepsilon_{ss} \end{pmatrix},$$

where $K = \rho^2\varepsilon_{\rho\rho} - 2\rho s\varepsilon_{\rho s} + s^2\varepsilon_{ss} + 2\rho\varepsilon_\rho$.

16. Using the result of the previous problem, prove that $e = \rho(\varepsilon + u^2/2)$ is convex in the variables $\mathbf{u} = (\rho, \rho u, \rho s)^T$ if and only if $E(\tau, s) = \varepsilon(1/\tau, s)$ is convex in the variables (τ, s).

17. Using the Godunov–Friedrichs–Lax theorem, write the system of conservation laws of gas dynamics

$$\rho_t + (\rho u)_x = 0,$$

$$(\rho u)_t + (p + \rho u^2)_x = 0,$$

$$(\rho s)_t + (\rho u s)_x = 0,$$

$$p = p(\rho, s)$$

as the symmetric t-hyperbolic Friedrichs system

$$A(\mathbf{v})\mathbf{v}_t + B(\mathbf{v})\mathbf{v}_x = 0$$

where $A = A^T > 0$ and $B = B^T$, by using the law of conservation of energy with the function $e = \rho(\varepsilon + \frac{1}{2}u^2)$ (here, ε is the internal energy).
Answer:

$$A = (e^*(\mathbf{v}))'' = (e(\mathbf{u}))'')^{-1}, \quad B = (ue^*(\mathbf{v}))'',$$

where

$$e^*(\mathbf{v}) = p, \quad \mathbf{v} = \left(\varepsilon + \frac{p}{\rho} - Ts - \frac{1}{2}u^2, u, T\right)^T.$$

18. Show that for any conservation law $\partial_t P + \partial_x Q = 0$ of the strictly hyperbolic system (1.22) written in Riemann invariants $\mathbf{r} = (r_1, \ldots, r_n)^T$ the function $P(r_1, \ldots, r_n)$ satisfies the system of linear equations

$$\frac{\partial^2 P}{\partial r_i \partial r_j} = \frac{1}{c_i - c_j}\left(\frac{\partial c_j}{\partial r_i}\frac{\partial P}{\partial r_j} - \frac{\partial c_i}{\partial r_j}\frac{\partial P}{\partial r_i}\right) \quad (i, j = 1, \ldots, n; \ i \neq j).$$

19. Find all conservation laws $\partial_t P(r, l) + \partial_x Q(r, l) = 0$ of the system in Riemann invariants

$$r_t + lr_x = 0,$$

$$l_t + rl_x = 0$$

(the Chaplygin isentropic gas with polytropic exponent $\gamma = -1$).
Answer: $P(r, l) = \frac{f(r) - g(l)}{r - l}$, where f and g are arbitrary smooth functions.

20. Find out whether there are conservation laws

$$\partial_t P(r, l, s) + \partial_x Q(r, l, s) = 0$$

for the system of equations

$$r_t + lr_x = 0,$$
$$l_t + sl_x = 0,$$
$$s_t + rs_x = 0.$$

Answer: Do not exist.

21. Find jumps of the derivative ρ_x through the lines of weak discontinuity of the solution to the Cauchy problem for the system of equations

$$\rho_t + u\rho_x + \rho u_x = 0,$$
$$u_t + uu_x + \rho\rho_x = 0$$

with the initial data

$$\rho(x,0) = \rho_0, \quad u(x,0) = \begin{cases} 0, & x \leqslant 0, \\ kx, & x > 0, \end{cases} \quad k, \rho_0 = \text{const} > 0.$$

Answer: $x = \pm\rho_0 t$: $[\rho_x] = \mp\frac{k}{2(1+kt)}$.

22. Show that discontinuities of the second order derivatives of a solution to a hyperbolic system that, together with the first order derivatives, is continuous can propagate only along characteristics.

23. Prove that the velocity of propagation of weak kinematic shock waves in a continuous medium with density ρ is expressed as

$$D = \frac{1}{2}(c_1 + c_2) + O([\rho]^2),$$

where c_i is the limit value of the characteristic velocity on the discontinuity line. Verify that this equality is exact for kinematic waves with a quadratic function $q = Q(\rho)$ of ρ.

24. Show that $D_1 < D < D_2$ for the velocity D of the kinematic shock wave caused by merging two shock waves with velocities $D_1 < D_2$ if the interaction of waves is described by the law of conservation $\partial_t \rho + \partial_x Q(\rho) = 0$, where Q is a convex function ($Q''(\rho) > 0$).

25. At point $x = 0$ at time $t = 0$, a kinematic shock wave travelling with constant velocity D_1 catches up a shock wave travelling with velocity D_0 over a constant state $\rho_0 > 0$. It is known that, after the wave interaction, the density ρ doubles and the process is described by the conservation law $\partial_t \rho + c_0 \partial_x(\rho^3/\rho_0^2) = 0$, $c_0 = \text{const} > 0$. Find the density $\rho(x,t)$ for $t > 0$. What are the velocities of all shock waves participated in the motion?

Answer: $\rho(x,t) = \rho_0(x > Dt)$, $\rho(x,t) = 4\rho_0(x < Dt)$, $D = 3D_0$, $D_1 = 4D_0$, $D_0 = 7c_0$.

26. A shock wave described by the conservation law $\partial_t u + \partial_x(u^2/2) = 0$ propagates with constant velocity $D = 2u_0 > 0$ in a fluid over the state $u = u_0$ and, at time $t = 1$, catches up the trailing edge of the centered wave $u = x/t$ $(u_0 t < x < 2u_0 t)$ travelling to the right in a fluid over the state $u = 2u_0$. Find the trajectory of the shock wave in the (x, t)-plane before and after the interaction with the centered wave. Find out whether the shock wave catches the leading edge of the centered wave.

Answer:

$$x = \begin{cases} u_0(2t - 1), & t \leqslant 1, \\ u_0(3t - 2\sqrt{t}), & 1 \leqslant t \leqslant 4, \\ u_0\left(\frac{5}{2}t - 2\right), & t \geqslant 4. \end{cases}$$

It catches up.

27. Consider kinematic waves in a traffic flow with the quadratic dependence $q = Q(\rho)$ of the flow on the vehicle density $\rho \in [0, \rho_*]$:

$$Q(\rho) = 4q_m \frac{\rho}{\rho_*}\left(1 - \frac{\rho}{\rho_*}\right), \qquad q_m = \text{const} > 0.$$

Verify the condition (1.12) for $Q(\rho)$. What is the maximal value $u_m = \max\limits_{0 \leqslant \rho \leqslant \rho_*} u(\rho)$ of the vehicle velocity $u(\rho) = Q(\rho)/\rho$? Find the dependence of the flow $q = q(u)$ on the velocity $u \in [0, u_m]$. For what values of u is the maximum of the flow $q_m = \max\limits_{0 \leqslant u \leqslant u_m} q(u)$ attained? What are the extremal values of the characteristic velocity $c(\rho) = Q'(\rho)$ on the interval $[0, \rho_*]$? Derive an expression in the form $c = c(q)$ for $q = Q(\rho)$ on the interval $[0, \rho_*/2]$.

Answer:

$$u_m = \frac{4q_m}{\rho_*}, \qquad q(u) = \rho_* u\left(1 - \frac{u}{u_m}\right), \qquad q_m = q(u)\big|_{u = u_m/2},$$

$$\max\limits_{0 \leqslant \rho \leqslant \rho_*} c(\rho) = c(\rho)\big|_{\rho=0} = u_m, \qquad \min\limits_{0 \leqslant \rho \leqslant \rho_*} c(\rho) = c(\rho)\big|_{\rho=\rho_*} = -u_m,$$

$$c(q) = u_m\sqrt{1 - \frac{q}{q_m}}.$$

28. Prove that, under the conditions (1.12), all continuous perturbations propagate only upstream in the traffic flow: $u(\rho) > c(\rho)$ for $0 < \rho \leqslant \rho_*$, where $u(\rho) = Q(\rho)/\rho$ is the vehicle velocity and $c(\rho) = Q'(\rho)$ is the characteristic velocity.

29. Prove that the propagation velocity of the kinematic shock wave $D(\rho) = (Q(\rho) - Q(\rho_0))/(\rho - \rho_0)$, regarded as a function of state ρ with a fixed state ρ_0

on the other side of the front, satisfies the relation

$$D'(\rho) = \frac{1}{(\rho - \rho_0)^2} \int_{\rho_0}^{\rho} (\varrho - \rho_0) Q''(\varrho) d\varrho.$$

Show that $D(\rho)$ is a monotone decreasing function if $Q''(\rho) < 0$.

30. The traffic flow moves with velocity u_0 and density ρ_0 along a street where the traffic light turns red at time $t = 0$ and is kept during the time T_k. Using the kinematic wave equations with flow

$$Q(\rho) = 4q_m \frac{\rho}{\rho_*} \cdot \left(1 - \frac{\rho}{\rho_*}\right) \quad (0 < \rho_0 < \rho_*, \ u_0 = Q(\rho_0)/\rho_0),$$

describe the traffic motion in a neighborhood of the traffic light for $t > 0$:

a. Find the trajectory OAB of the deceleration "shock wave" propagating upstream in the traffic flow (cf. Fig. 1.6).
b. Determine the time t_B when the wave OAB comes to a point B (if $T_{green} > t_B - T_k$, then no traffic bottleneck occurs at the traffic light working in the green light mode during the time T_{green}).
c. Indicate the time t_C when the vehicles which have been halted because of a red traffic light catch up the flow of the vehicles which have passed by the traffic light before the time $t = 0$.

Answer:

a.

$$OA: \quad x = \frac{1}{2}(c_0 - u_m)t \quad \left(c_0 = Q'(\rho_0), u_m = \frac{4q_m}{\rho_*}\right),$$

$$AB: \quad x = c_0(t - T_k) - \sqrt{(u_m^2 - c_0^2)T_k(t - T_k)}.$$

b.

$$t_B = \frac{q_m}{q_m - q_0} T_k \quad (q_0 = \rho_0 u_0).$$

c.

$$t_C = \frac{u_m}{u_m - u_0} T_k.$$

31. Because of a road accident, the traffic flow q_0 moved with density ρ_0 decreases up to $q_1 < q_0$ at the accident place during the time interval T. Using the kinematic wave equation with the same function $Q(\rho)$ as in the previous problem and assuming that $0 < q_0 = Q(\rho_0) < q_m$ and $0 < \rho_0 < \rho_*/2$,

find the maximal distance l from the accidence place, where the road accident affects the transport delay.

Answer:

$$l = \frac{\sqrt{q_m}(q_0 - q_1)}{\rho_* \sqrt{q_m - q_0}} T.$$

32. Show that a simple wave for the equations of one-dimensional isentropic flows of a polytropic gas propagating through the state at rest with density ρ_0 and sound speed c_0 is a kinematic wave. Find the dependence $q = Q(\rho)$ of the flow $q = \rho u$ on the density ρ.

Answer:

$$Q(\rho) = \pm \frac{2c_0}{\gamma - 1} \rho \left(1 - \left(\frac{\rho}{\rho_0}\right)^{\frac{\gamma - 1}{2}}\right).$$

33. The process of filtration of a fluid in a porous medium is governed by the Buckley–Leverett equation

$$\partial_t s + \partial_x Q(s) = 0, \quad Q(s) = 3s^2 - 2s^3,$$

where $0 \leqslant s(x, t) \leqslant 1$ is the saturation of pores by the fluid. Show that the Riemann problem with the initial data $s(x, 0) = 0$ for $x > 0$ and $s(x, 0) = 1$ for $x < 0$ has at least two self-similar solutions. One of the solutions is a combination of a shock and the adjacent centered wave (a stable solution):

$$s(x, t) = 1 \ (x < 0), \quad s(x, t) = 0 \ \left(\frac{x}{t} > \frac{9}{8}\right),$$

$$s(x, t) = \frac{1}{2} + \frac{1}{2}\sqrt{1 - \frac{2x}{3t}} \ \left(0 < \frac{x}{t} < \frac{9}{8}\right),$$

whereas the other is a piecewise constant (an unstable solution):

$$s(x, t) = 1 \ \left(\frac{x}{t} < 1\right), \quad s(x, t) = 0 \ \left(\frac{x}{t} > 1\right).$$

Is the function $Q(s)$ convex on the interval $0 \leqslant s(x, t) \leqslant 1$?

34. Construct a travelling wave type solution $u(x, t) = U(x - Dt)$ for the Burgers equation

$$u_t + u u_x = \nu u_{xx} \quad (\nu = \text{const} > 0)$$

such that $u \to u_1$ as $x \to +\infty$ and $u \to u_2$ as $x \to -\infty$, where u_1 and u_2 are constants ($u_1 \neq u_2$). Find asymptotics of the solution as $\nu \to 0$.

Answer:

$$u(x,t) = u_1 + \frac{u_2 - u_1}{1 + \exp\left\{\frac{u_2 - u_1}{2v}(x - Dt)\right\}}, \quad D = \frac{1}{2}(u_1 + u_2),$$

$$\lim_{v \to 0} u(x,t) = \begin{cases} u_1, & x > Dt, \\ u_2, & x < Dt. \end{cases}$$

35. Show that the function $u = -2vv_x/v$ is a solution to the Burgers equation if the function v satisfies the heat equation

$$v_t = vv_{xx}$$

(the Cole–Hopf transform). What solution $v_*(x,t)$ to the heat equation yields a self-similar solution to the Burgers equation $u_*(x,t) = x/t$ under this transform?
Answer:

$$v_*(x,t) = \frac{1}{\sqrt{t}} e^{-x^2/(4vt)}.$$

36. Verify that the solution to the Burgers equation in Problem 34 is the Cole–Hopf transform of the sum $v = v_1 + v_2$ of two travelling wave type solutions to the heat equation of the form

$$v_j(x,t) = \exp\left\{-\frac{u_j}{2v}\left(x - \frac{u_j}{2}t\right)\right\} \quad (j = 1, 2). \qquad (1.24)$$

37. Consider the solution $u(x,t)$ to the Burgers equation which is the Cole–Hopf transform of the sum $v(x,t) = v_1(x,t) + v_2(x,t) + v_3(x,t)$ of three solutions to the heat equation of the form (1.24) with parameters $u_3 > u_2 > u_1$. Find the asymptotics of u as $v \to 0$.
Answer:

$$\lim_{v \to 0} u(x,t) = \begin{cases} u_1, & x > D_1t, \\ u_2, & D_2t < x < D_1t, \\ u_3, & x < D_3t, \end{cases} \quad \text{for } t < 0,$$

$$\lim_{v \to 0} u(x,t) = \begin{cases} u_1, & x > D_3t, \\ u_3, & x < D_3t, \end{cases} \quad \text{for } t \geq 0,$$

where

$$D_1 = \frac{1}{2}(u_1 + u_2), \quad D_2 = \frac{1}{2}(u_2 + u_3), \quad D_3 = \frac{1}{2}(u_1 + u_3).$$

38. Prove that for the system of equations

$$\mathbf{u}_t + A(\mathbf{u})\mathbf{u}_x = 0,$$

$$\mathbf{u} = (u_1, u_2, u_3)^T,$$

there exists a Riemann invariant constant along the characteristic $dx/dt = c$ if and only if $\mathbf{l} \cdot \mathrm{curl}\,\mathbf{l} = 0$, where $\mathbf{l}(\mathbf{u})$ is a left eigenvector of the 3×3-matrix A corresponding to the eigenvalue c.

39. For what characteristics do Riemann invariants exist for the equations of one-dimensional motion of a polytropic gas

$$\rho_t + u\rho_x + \rho u_x = 0,$$

$$u_t + uu_x + \frac{1}{\rho}p_x = 0,$$

$$p_t + up_x + \gamma p u_x = 0.$$

Answer:

$$\frac{dx}{dt} = u(x,t): \quad p\rho^{-\gamma} = \mathrm{const}$$

(the entropy is constant on the contact characteristic).

40. For what dependence of the sound speed $c = c(\rho, p)$ on the density ρ and pressure p do Riemann invariants $r_\pm$ exist and are conserved on the sound characteristics $dx/dt = u \pm c$ of the system of equations of one-dimensional gas dynamics

$$\rho_t + u\rho_x + \rho u_x = 0,$$

$$u_t + uu_x + \frac{1}{\rho}p_x = 0,$$

$$p_t + up_x + \rho c^2 u_x = 0.$$

Answer:

$$c(\rho, p) = \frac{a(p)}{\rho}, \quad r_\pm = u \pm \int \frac{dp}{a(p)},$$

where $a(p)$ is an arbitrary function.

41. Consider the hyperbolic system of conservation laws

$$\partial_t \mathbf{u} + \partial_x \boldsymbol{\psi}(\mathbf{u}) = 0. \tag{1.25}$$

Prove that for any branch of the shock adiabat the tangent vector to this branch at the adiabat center $\mathbf{u} = \mathbf{u}_0$ is a right eigenvector of the matrix $A(\mathbf{u}_0) = \boldsymbol{\psi}'(\mathbf{u}_0)$.

42. We look for a self-similar solution $\mathbf{u} = \mathbf{u}(\xi)$, $\xi = x/t$, to the strictly hyperbolic system of equations

$$\partial_t \mathbf{u} + A(\mathbf{u})\partial_x \mathbf{u} = 0.$$

Show that a necessary conditions for the existence of such a solution is the condition $\mathbf{r}(\mathbf{u}) \cdot \nabla_{\mathbf{u}} c(\mathbf{u}) \neq 0$ for at least one eigenvalue $c(\mathbf{u})$ of the matrix $A(\mathbf{u})$ (here, $\mathbf{r}$ is a right eigenvector of the matrix A corresponding to the eigenvalue c). Show that if the characteristic field $c(\mathbf{u})$ satisfies this condition, called the *genuine nonlinearity in the sense of Lax*, then the solution $\mathbf{u}(\xi)$ is obtained by integrating the system of ordinary differential equations

$$\frac{d\mathbf{u}}{d\xi} = \frac{\mathbf{r}(\mathbf{u})}{\mathbf{r}(\mathbf{u}) \cdot \nabla_{\mathbf{u}} c(\mathbf{u})}.$$

43. The characteristic field $c(\mathbf{u})$ is said to be *linearly degenerate in the sense of Lax* if $\mathbf{r}(\mathbf{u}) \cdot \nabla_{\mathbf{u}} c(\mathbf{u}) \equiv 0$ for a right edigenvector $\mathbf{r}$ corresponding to the eigenvalue c. Show that if $c(\mathbf{u})$ is a simple eigenvalue of the matrix $A(\mathbf{u}) = \boldsymbol{\psi}'(\mathbf{u})$ of the hyperbolic conservation law system (1.25), then the shock adiabat branch $D(\mathbf{u} - \mathbf{u}_0) = \boldsymbol{\psi}(\mathbf{u}) - \boldsymbol{\psi}(\mathbf{u}_0)$ tangent to the vector $\mathbf{r}(\mathbf{u}_0)$ at the point $\mathbf{u} = \mathbf{u}_0$ coincides with the integral curve $\mathbf{u} = \mathbf{u}(s)$ of the system of ordinary differential equations

$$\frac{d\mathbf{u}}{ds} = \mathbf{r}(\mathbf{u}),$$

$$\mathbf{u}|_{s=0} = \mathbf{u}_0.$$

Moreover, $D = c(\mathbf{u})$ for the shock wave velocity.

44. Show that for the system of gas dynamics

$$\rho_t + u\rho_x + \rho u_x = 0,$$

$$u_t + uu_x + \frac{1}{\rho}p_x = 0,$$

$$s_t + us_x = 0,$$

$$p = p(\rho, s)$$

the sound characteristics $dx/dt = u \pm c$ (here, $c = \sqrt{p_\rho(\rho, s)}$ is the sound speed) satisfy the condition of genuine nonlinearity in the sense of Lax if the equation of state $p = g(\tau, s)$, $\tau = 1/\rho$, is such that $g_{\tau\tau}(\tau, s) > 0$. Show that the contact characteristics $dx/dt = u$ are linearly degenerate in the sense of Lax.

45. Consider the system of conservation laws of mass, momentum, and energy of the one-dimensional motion of an polytropic gas

$$\rho_t + (\rho u)_x = 0,$$

$$(\rho u)_t + (\rho u^2 + p)_x = 0,$$

$$\left(\frac{1}{\gamma - 1}p + \frac{1}{2}\rho u^2\right)_t + \left(\frac{1}{2}\rho u^3 + \frac{\gamma}{\gamma - 1}pu\right)_x = 0.$$

Write the shock adiabat equation for a polytropic gas in the space (ρ, u, p) with the origin (ρ_0, u_0, p_0) in the parametric form with the Mach number $M = |u_0 - D|/c_0$ for a parameter, where D is the shock wave velocity and $c_0 = \sqrt{\gamma p_0/\rho_0}$.

Answer:

$$\rho = \rho_0 \left\{ 1 + \frac{2(M^2 - 1)}{(\gamma - 1)M^2 + 2} \right\}, \quad u = u_0 \pm \frac{2|M^2 - 1|}{(\gamma + 1)M}c_0,$$

$$p = p_0 \left\{ 1 + \frac{2\gamma}{\gamma + 1}(M^2 - 1) \right\}.$$

46. Find characteristics of the system of linear elasticity

$$\frac{\partial u_i}{\partial t} = \frac{\partial v_i}{\partial x},$$

$$\rho_0 \frac{\partial v_i}{\partial t} = \frac{\partial}{\partial x}\frac{\partial \Phi}{\partial u_i} \quad (i = 1, 2, 3)$$

with potential

$$\Phi(u_1, u_2, u_3) = \frac{1}{2}\mu(u_1^2 + u_2^2) + \frac{1}{2}(\lambda + 2\mu)u_3^2,$$

where $0 < \rho_0 = $ const is the density of the material in the nondeformed state, u_i are the strains, v_i are the velocities, λ and μ are the Lamé constants ($\lambda > 0, \mu > 0$). Is this system hyperbolic?

Answer:

$$\frac{dx}{dt} = c_j \quad (j = 1, \ldots, 6),$$

where

$$c_{1,2} = \sqrt{\frac{\mu}{\rho_0}}, \quad c_{3,4} = -\sqrt{\frac{\mu}{\rho_0}} \quad \text{(transverse waves)},$$

$$c_{5,6} = \pm \sqrt{\frac{\lambda + 2\mu}{\rho_0}} \quad \text{(longitudinal waves)}.$$

The system is hyperbolic.

47. Under the conditions of the previous problem, show that the linear equations of elasticity have a solution with a strain vector $\mathbf{u} = (u_1, u_2, u_3)^T$ of the form

$$\mathbf{u}(x, t) = f(x - c_1 t)\mathbf{a} + g(x - c_5 t)\mathbf{b}$$

where $c_1 = \sqrt{\mu/\rho_0}$, $c_5 = \sqrt{(\lambda + 2\mu)/\rho_0}$, $\mathbf{a}, \mathbf{b} \in \mathbb{R}^3$ are constant vectors such that $\mathbf{a} \cdot \mathbf{b} = 0$, and the functions f and g are not identically constants.

48. Find the characteristic velocities for the system of quasitransverse wave equations

$$\partial_t u_i = \partial_x v_i,$$

$$\rho_0 \partial_t v_i = \partial_x \Phi_{u_i},$$

where $i = 1, 2$, in an isotropic nonlinear medium with elastic potential

$$\Phi(u_1, u_2) = \frac{1}{2}\mu(u_1^2 + u_2^2) + \frac{1}{4}\kappa^2(u_1^2 + u_2^2)^2 \quad (\kappa = \text{const}).$$

Answer:

$$c_{1,2} = \pm\sqrt{\frac{\mu + \kappa^2(u_1^2 + u_2^2)}{\rho_0}},$$

$$c_{3,4} = \pm\sqrt{\frac{\mu + 3\kappa^2(u_1^2 + u_2^2)}{\rho_0}}.$$

49. Find characteristics and Riemann invariants of the system of equations of longitudinal nonlinear elastic waves in a rod

$$u_t = v_x,$$

$$\rho_0 v_t = \sigma_x,$$

where u is the strain and $\sigma = \sigma(u)$ is the stress ($\sigma'(u) > 0$).

Answer:

$$\frac{dx}{dt} = \pm\sqrt{\frac{\sigma'(u)}{\rho_0}},$$

$$r_\pm(u, v) = v \mp \int\limits_0^u \sqrt{\frac{\sigma'(\xi)}{\rho_0}}\, d\xi.$$

50. Show that the shock adiabat (1.10) of the system of equations of nonlinear elastic waves in a rod has the second order tangency at the center (u_0, v_0) with the level lines of the Riemann invariants $r_\pm(u, v) = r_\pm(u_0, v_0)$.

51. Find characteristics and Riemann invariants of the equilibrium equations of an ideal rigid-plastic body for plane deformations

$$\frac{\partial\sigma}{\partial x} - 2k\left(\cos 2\theta\frac{\partial\theta}{\partial x} + \sin 2\theta\frac{\partial\theta}{\partial y}\right) = 0,$$

$$\frac{\partial\sigma}{\partial y} - 2k\left(\sin 2\theta\frac{\partial\theta}{\partial x} - \cos 2\theta\frac{\partial\theta}{\partial y}\right) = 0,$$

where $\sigma(x, y) = (\sigma_{11} + \sigma_{22})/2$ is the mean stress, $k = \text{const}$ is the yield limit under shear (the maximal tangent stress), $\theta(x, y)$ is the angle of slope of the line of maximal tangent stress at the point (x, y).

Answer:

$$\frac{dy}{dx} = \tan\theta : \qquad \sigma - 2k\theta = \text{const}$$

$$\frac{dy}{dx} = -\cotan\theta : \qquad \sigma + 2k\theta = \text{const} .$$

52. Consider the equation of plane stress state of a rigid-plastic material under the von Mises yield criterion of the form

$$\left(\sqrt{3}\sin\omega\cos 2\varphi - \cos\omega\right)\frac{\partial\omega}{\partial x} + \sqrt{3}\sin\omega\sin 2\varphi\frac{\partial\omega}{\partial y} - 2\sin\omega\frac{\partial\varphi}{\partial y} = 0,$$

$$\sqrt{3}\sin\omega\sin 2\varphi\frac{\partial\omega}{\partial x} - \left(\sqrt{3}\sin\omega\cos 2\varphi + \cos\omega\right)\frac{\partial\omega}{\partial y} + 2\sin\omega\frac{\partial\varphi}{\partial x} = 0,$$

where the function $\omega(x, y)$ is connected with the principal stresses σ_1 and σ_2 by the equalities

$$\sigma_1 = 2k\cos\left(\omega - \frac{\pi}{6}\right),$$

$$\sigma_2 = 2k\cos\left(\omega + \frac{\pi}{6}\right),$$

where $k = $ const is the yield limit, and $\varphi(x, y)$ is the angle between the first principal direction of the stress tensor and the Ox-axis. Find characteristics and Riemann invariants in the domain where the system is hyperbolic.

Answer:

$$\frac{dy}{dx} = \frac{\sqrt{3}\sin\omega \sin 2\varphi \pm \sqrt{3 - 4\cos^2\omega}}{\sqrt{3}\sin\omega \cos 2\varphi - \cos\omega} :$$

$$\varphi \mp \int_{\pi/6}^{\omega} \frac{\sqrt{3 - 4\cos^2 s}}{2\sin s}\, ds = \text{const}$$

$$\left(\frac{\pi}{6} < \omega < \frac{5\pi}{6}, \quad \frac{7\pi}{6} < \omega < \frac{11\pi}{6}\right).$$

53. Derive the Hamilton–Jacobi equation for the characteristic surfaces of the system of two-dimensional equations of linear elasticity

$$\rho_0 \frac{\partial v_1}{\partial t} = \frac{\partial \sigma_{11}}{\partial x} + \frac{\partial \sigma_{12}}{\partial y},$$

$$\rho_0 \frac{\partial v_2}{\partial t} = \frac{\partial \sigma_{21}}{\partial x} + \frac{\partial \sigma_{22}}{\partial y},$$

$$\frac{\partial \sigma_{11}}{\partial t} = (\lambda + 2\mu)\frac{\partial v_1}{\partial x} + \lambda\frac{\partial v_2}{\partial y},$$

$$\frac{\partial \sigma_{22}}{\partial t} = \lambda\frac{\partial v_1}{\partial x} + (\lambda + 2\mu)\frac{\partial v_2}{\partial y},$$

$$\frac{\partial \sigma_{12}}{\partial t} = \mu\left(\frac{\partial v_1}{\partial y} + \frac{\partial v_2}{\partial x}\right) \quad (\sigma_{12} = \sigma_{21}),$$

where $\mathbf{v} = (v_1, v_2)^T$ is the velocity vector and σ_{ij} are components of the stress tensor.

Answer:

$$\varphi_t = 0, \quad \varphi_t = \pm\sqrt{\frac{\mu}{\rho_0}}\sqrt{\varphi_x^2 + \varphi_y^2}, \quad \varphi_t = \pm\sqrt{\frac{\lambda + 2\mu}{\rho_0}}\sqrt{\varphi_x^2 + \varphi_y^2}.$$

54. We consider the chemical adsorption equations (1.4) for a two-component mixture with the Langmuir isotherm

$$\mathbf{f}(\mathbf{u}) = \frac{1}{p}(\Gamma_1 u_1, \Gamma_2 u_2)^T, \quad p(u_1, u_2) = 1 + \Gamma_1 u_1 + \Gamma_2 u_2,$$

where Γ_k are the Henry coefficients ($0 < \Gamma_1 < \Gamma_2$). Show that for $u_1 > 0$ and $u_2 > 0$ the following assertions hold.

a. $\lambda \in \mathbb{R}$ is an eigenvalue of the matrix $\mathbf{f}'(\mathbf{u})$ if and only if λ is a root of the equation

$$\frac{\Gamma_1^2 u_1}{\Gamma_1 - p\lambda} + \frac{\Gamma_2^2 u_2}{\Gamma_2 - p\lambda} = p. \qquad (1.26)$$

b. The eigenvalues λ_j of the matrix $\mathbf{f}'(\mathbf{u})$ are real and distinct: $0 < \lambda_1 < \Gamma_1/p < \lambda_2 < \Gamma_2/p$ (the strict hyperbolicity of the Langmuir adsorption equations).

55. Verify that the chemical adsorption equations (1.4) can be transformed to the form

$$\partial_\xi \mathbf{u} + \partial_\tau \mathbf{f}(\mathbf{u}) = 0$$

by making the change of independent variables $\tau = vt - x$, $\xi = x$. Show that for every eigenvalue λ of the matrix $\mathbf{f}'(\mathbf{u})$ given by Eq. (1.26) the function $r = p\lambda$ is a Riemann invariant ($\partial_\xi + \lambda \partial_\tau)r = 0$ for the characteristic $d\tau/d\xi = \lambda$.

56. Consider the system of equations written in Riemann invariants

$$\partial_\xi r_1 + \frac{r_1^2 r_2}{\Gamma_1 \Gamma_2} \partial_\tau r_1 = 0,$$

$$\partial_\xi r_2 + \frac{r_1 r_2^2}{\Gamma_1 \Gamma_2} \partial_\tau r_2 = 0.$$

Show that the functions

$$u_1 = \frac{\Gamma_2/\Gamma_1}{\Gamma_1 - \Gamma_2} \left(\frac{\Gamma_1}{r_1} - 1 \right) \left(\frac{\Gamma_1}{r_2} - 1 \right),$$

$$u_2 = \frac{\Gamma_1/\Gamma_2}{\Gamma_2 - \Gamma_1} \left(\frac{\Gamma_2}{r_1} - 1 \right) \left(\frac{\Gamma_2}{r_2} - 1 \right)$$

with constant $\Gamma_1 \neq \Gamma_2$ satisfy the system of the Langmuir adsorption equations for a two-component mixture

$$\partial_\xi u_1 + \partial_\tau \left(\frac{\Gamma_1 u_1}{1 + \Gamma_1 u_1 + \Gamma_2 u_2} \right) = 0,$$

$$\partial_\xi u_2 + \partial_\tau \left(\frac{\Gamma_2 u_2}{1 + \Gamma_1 u_1 + \Gamma_2 u_2} \right) = 0.$$

Remark The Langmuir adsorption equations are also reduced to a system in Riemann invariants in the general case of an *n*-component mixture.

57. Show that the laws of conservation of mass and momentum in gas dynamics

$$\rho_t + \text{div}(\rho\mathbf{u}) = 0,$$

$$(\rho\mathbf{u})_t + \text{div}(\rho\mathbf{u} \otimes \mathbf{u} + pI) = 0$$

imply the equation for the density

$$\rho_{tt} - c_0^2 \triangle\rho = \sum_{i,j=1}^{3} \frac{\partial^2 T_{ij}}{\partial x_i \partial x_j},$$

where c_0 is the sound speed in a gas at rest, $T_{ij} = \rho u_i u_j + (p - c_0^2\rho)\delta_{ij}$ ($i,j = 1, 2, 3$) are components of the acoustic stress tensor, $\mathbf{u} = (u_1, u_2, u_3)^T$ is the velocity vector, p is the pressure, and δ_{ij} is the Kronecker symbol.

58. Show that the potential φ of the velocity field $\mathbf{u} = \nabla\varphi$ of a vortex-free isentropic gas flow satisfies the equation

$$d_t\left(\varphi_t + \frac{1}{2}|\nabla\varphi|^2\right) = c^2\triangle\varphi, \quad (d_t = \partial_t + \nabla\varphi \cdot \nabla),$$

where c is the sound speed.

59. Show that for the system of linear acoustic equations in a gas flow of density $\rho_0 = \text{const}$ moving with a given constant speed $\mathbf{u}_0$,

$$\rho_0\, d_t\mathbf{u} + \nabla p = 0,$$

$$d_t p + \rho_0 c_0^2 \,\text{div}\,\mathbf{u} = 0,$$

where $d_t = \partial_t + \mathbf{u}_0 \cdot \nabla$, the following integral law of conservation of energy holds:

$$\frac{1}{2}\frac{d}{dt}\iiint_\Omega \left(\rho_0|\mathbf{u}|^2 + \frac{p^2}{\rho_0 c_0^2}\right) d\Omega + \frac{1}{2}\iint_S \left(\rho_0|\mathbf{u}|^2 + \frac{p^2}{\rho_0 c_0^2}\right)\mathbf{u}_0 \cdot \mathbf{n}\, dS$$

$$+ \iint_S p\mathbf{u} \cdot \mathbf{n}\, dS = 0,$$

where $\Omega \subset \mathbb{R}^3$ is an arbitrary domain with piecewise smooth boundary S and $\mathbf{n}$ is the outward unit normal to S.

60. For the acoustic equations in a gas moving with the constant speed $\mathbf{u}_0 = (u_0, 0, 0)^T$ find sound characteristics and corresponding bicharacteristics provided that each of these characteristic surfaces at $t = 0$ is the sphere $|\mathbf{x}| = R$.
    Answer: $(x + u_0 t)^2 + y^2 + z^2 = (R \pm c_0 t)^2$ are characteristics,
    $x = \left(1 \pm \frac{c_0 t}{R}\right) x_0 - u_0 t$, $y = \left(1 \pm \frac{c_0 t}{R}\right) y_0$, and $z = \left(1 \pm \frac{c_0 t}{R}\right) z_0$ are bicharacteristics.

61. Let the dependence of $p(\mathbf{x}, t)$ on the variables $\mathbf{x} = (x, y, z)^T$ and t be implicitly given by

$$p = f(\mathbf{x} \cdot \mathbf{k}(p) + c_0 t |\mathbf{k}(p)|),$$

where $f, \mathbf{k} \in C^2$. Show that p is a solution to the wave equation

$$p_{tt} = c_0^2 \Delta p$$

(the Smirnov–Sobolev functional-invariant solution).

62. Find a general form of a spherical wave type solution $p = p(|\mathbf{x}|, t)$ for the wave equation.
    Answer: $p = \frac{1}{|\mathbf{x}|}(f(|\mathbf{x}| - c_0 t) + g(|\mathbf{x}| + c_0 t))$, where f and g are arbitrary smooth functions.

63. The acoustic radiation by a ball of radius r_0 that pulsates according to the harmonic law with frequency ω is described by the boundary condition $u_r(\mathbf{x}, t) = U \cos \omega t$ ($U = \text{const} \neq 0$) for $|\mathbf{x}| = r_0$ for the radial component u_r of the gas velocity $\mathbf{u}$. Looking for a solution to the acoustic equation in the form of a spherical wave outgoing from the sound source (i.e., depending on $|\mathbf{x}| - c_0 t$), find the pressure field p in the domain $|\mathbf{x}| > r_0$.
    Answer:

$$p(\mathbf{x}, t) = \text{Re}\left\{ \frac{a}{|\mathbf{x}|} e^{i\omega(t - |\mathbf{x}|/c_0)} \right\}, \qquad a = \frac{i\rho_0 c_0 r_0 m U}{1 + im} e^{im} \qquad \left(m = \frac{\omega r_0}{c_0}\right).$$

64. Find the acoustic energy flux I through the sphere $S_R : |\mathbf{x}| = R$,

$$I = \int_{S_R} p \mathbf{u} \cdot \mathbf{n} \, dS,$$

for the sound source of frequency ω in a gas at rest that creates the pressure distribution $p(\mathbf{x}, t) = a|\mathbf{x}|^{-1} \cos \omega(t - |\mathbf{x}|/c_0)$.
    Answer:

$$I = \frac{4\pi a^2}{\rho_0 c_0}\left(\cos^2[\omega(t - R/c_0)] + \frac{c_0}{2\omega R} \sin[2\omega(t - R/c_0)] \right).$$

65. The Maxwell equations governing the propagation of electromagnetic waves in a medium with dielectric permeability ε and magnetic permeability μ have the form

$$\mathbf{b}_t + \operatorname{curl} \mathbf{e} = 0,$$
$$\mu\varepsilon\mathbf{e}_t - \operatorname{curl} \mathbf{b} = 0,$$
$$\operatorname{div} \mathbf{e} = 0,$$
$$\operatorname{div} \mathbf{b} = 0.$$

Show that the electric-field vector $\mathbf{e}$ and the magnetic induction vector $\mathbf{b}$ satisfy the wave equation. Determine the wave propagation velocity.

66. Show that for the Maxwell equations the energy integral identity holds:

$$\frac{1}{2}\frac{d}{dt}\iiint_\Omega (\mu\varepsilon|\mathbf{e}|^2 + |\mathbf{b}|^2)d\Omega + \iint_\Gamma \mathbf{s}\cdot\mathbf{n}\,d\Gamma = 0,$$

where $\mathbf{s} = \mathbf{e}\times\mathbf{b}$ is the Umov–Poynting vector (the energy flux vector), $\Omega \subset \mathbb{R}^3$ is an arbitrary domain with piecewise smooth boundary $\Gamma = \partial\Omega$, and $\mathbf{n}$ is the outward unit normal vector.

67. Find characteristics of the equations of magnetic gas dynamics

$$\rho_t + u\rho_x + \rho u_x = 0,$$
$$\rho(u_t + uu_x) + p_x = jb,$$
$$p_t + up_x + \gamma pu_x = \sigma^{-1}(\gamma - 1)j^2,$$
$$b_t + e_x = 0,$$
$$\mu\varepsilon e_t + b_x + \mu j = 0,$$

where $j = \sigma(e - ub)$ is the electrical current in the gas and $\sigma = \text{const}$ is the electrical conductivity.

Answer:

$$c_1 = u, \quad c_{2,3} = u \pm \sqrt{\frac{\gamma p}{\rho}}, \quad c_{4,5} = \pm\frac{1}{\sqrt{\mu\varepsilon}}.$$

68. Find the characteristics of equations of one-dimensional motion of an ideal gas
with infinite conductivity

$$\rho_t + u\rho_x + \rho u_x = 0,$$

$$b_t + ub_x + bu_x = 0,$$

$$\rho(u_t + uu_x) + p_x + \frac{1}{\mu}bb_x = 0,$$

$$p_t + up_x + \gamma pu_x = 0.$$

Answer:

$$c_{1,2} = u, \quad c_{3,4} = u \pm \sqrt{\frac{\gamma p}{\rho} + \frac{b^2}{\mu\rho}}.$$

69. Construct a simple centered wave type solution with density ρ playing the
role of a parameter of a simple wave of the equation of a gas with polytropic
exponent $\gamma = 2$ having infinite conductivity.
Answer:

$$b = A\rho, \quad u = \pm k\frac{\sqrt{\rho}}{2} + B, \quad p = \left(\frac{k^2}{32} - \frac{A^2}{2\mu}\right)\rho^2, \quad \rho = \left(\frac{x}{t} - B\right)^2\frac{16}{9k^2},$$

where A, B and k are constants.

70. Show that for each root $\tau = H(\xi; x, t)$ of the characteristic equation

$$\det\left(\tau I + \sum_{i=1}^{3}\xi_i A^i\right) = 0$$

of the system (1.14) the Hamiltonian H satisfies the Euler identity

$$H = \sum_{i=1}^{3}\xi_i\partial_{\xi_i}H.$$

71. Find a solution to the Cauchy problem for the Hamilton–Jacobi equation with
initial data of a plane wave type

$$\varphi_t = H(\nabla_{\mathbf{x}}\varphi), \quad \varphi(\mathbf{x}, 0) = \mathbf{k}\cdot\mathbf{x},$$

where $\mathbf{k} \in \mathbb{R}^3$ is a constant vector, under the condition that the Hamiltonian H is
a positive–homogeneous function of the first degree: $H(\lambda\mathbf{p}) = \lambda H(\mathbf{p})$ $(\lambda > 0)$.
Answer: $\varphi(\mathbf{x}, t) = \mathbf{k}\cdot\mathbf{x} + H(\mathbf{k})t$.

72. It is known that the wave front $\Gamma(t)$ is a characteristic of the system of equations

$$u_t + x v_y = 0,$$
$$v_t + y u_x = 0$$

and is given by the equation $\Gamma_0 : \varphi_0(x, y) = 0$ at $t = 0$. Find the trajectory of the ray outgoing from the point $(x_0, y_0) \in \Gamma_0$ in the (x, y)-plane. Find the location of the front at time $t > 0$ provided that, at $t = 0$, it has the shape of a hyperbola $xy = 1$ and perturbations propagate to the domain $xy < 1$.
Answer:

$$(x/x_0)^{x_0 p_0} = (y/y_0)^{y_0 q_0}, \quad p_0 = \partial_x \varphi_0(x_0, y_0),$$
$$q_0 = \partial_y \varphi_0(x_0, y_0), \quad xy = e^{-t}.$$

Chapter 2
Dispersive Waves

2.1 Dispersion Relation

We consider wave processes governed by the system of linear partial differential equations with constant coefficients

$$\sum_{s=0}^{n}\sum_{p=0}^{m} b_{sp}\partial_t^s \partial_x^p u(x,t) = 0, \tag{2.1}$$

where t is the time, $x \in \mathbb{R}$ is the spatial variable, and the coefficients b_{sp} and solution u can be complex-valued. A wave described by a complex-valued solution

$$u(x,t) = ae^{i(kx - \omega t)}, \tag{2.2}$$

is called an *elementary wave packet*. Here, a is the wave amplitude, k is the wavenumber, ω is the frequency, and $\theta = kx - \omega t$ is the wave phase. In the case of real parameters ($\operatorname{Im} k = 0$ and $\operatorname{Im}\omega = 0$), the wavenumber is equal to the number of waves at a segment of length 2π in the x-axis and the frequency is equal to the number of wave crests or troughs passing by a fixed observer during the time 2π. For such parameters we can introduce the *wavelength* $L = 2\pi/k$ and the *time period* $T = 2\pi/\omega$. Each constant value of the phase θ is carried with the velocity $c_p = \omega/k$, called the *phase velocity*. Taking into account this fact, we can regard a wave packet as a travelling wave $u(x,t) = a\exp ik(x - c_p t)$ propagating with the phase velocity.

Differentiating the function u defined by (2.2), we get

$$\partial_t u = -i\omega u,$$

$$\partial_x u = iku.$$

© Springer International Publishing AG 2017
S.L. Gavrilyuk et al., *Waves in Continuous Media*, Lecture Notes in Geosystems
Mathematics and Computing, DOI 10.1007/978-3-319-49277-3_2

Therefore, the system (2.1) have solutions in the form of wave packets with an amplitude $a \neq 0$ if and only if

$$D(\omega, k) = 0, \tag{2.3}$$

where

$$D(\omega, k) = \sum_{s=0}^{n} \sum_{p=0}^{m} b_{sp}(-i\omega)^s (ik)^p.$$

The equality (2.3) relating the frequency and the wavenumber is called the *dispersion relation*. Since $D(\omega, k)$ is a polynomial of degree n in ω, for a given k Eq. (2.3) has, in general, n complex roots $\omega_j = \omega_j(k)$ ($j = 1, \ldots, n$). The family of wave packets

$$u(x, t) = a \exp i(kx - \omega(k)t)$$

with an arbitrary wavenumber k and the frequency $\omega(k)$ generated by some fixed root of the dispersion relation is called a *wave mode*. The number of wave modes for the system (2.1) coincides with the order of this system with respect to the variable t. The dependence $c_p(k) = \omega(k)/k$ of the phase velocity on the wavenumber means that the profile of a wave consisting of several wave packets of a given mode with different k is deformed due to spreading of these packets running with different velocities. This phenomenon is called *dispersion* of waves. Respectively, a wave is *dispersive* if $\omega''(k) \neq 0$.

Example 2.1 Consider the telegraph equation

$$u_{tt} - c_0^2 u_{xx} + \alpha^2 u = 0, \tag{2.4}$$

where $c_0 \neq 0$ and α are real constants (this equation governs oscillations of electromagnetic waves along a conductor of large length, referred to as a transmission line). This equation yields the dispersion relation (2.3) with

$$D(\omega, k) = -\omega^2 + c_0^2 k^2 + \alpha^2$$

which generates two wave modes with the real frequencies

$$\omega_{\pm}(k) = \pm\sqrt{c_0^2 k^2 + \alpha^2}, \quad k \in \mathbb{R}.$$

For $k > 0$ the mode with frequency $\omega_+(k)$ describes waves travelling to the right along the Ox-axis, whereas the mode with frequency $\omega_-(k)$ describes waves propagating to the left. For $\alpha \neq 0$ we have $\omega''_{\pm}(k) \neq 0$, i.e., the waves are dispersive.

In the case $\alpha = 0$, Eq. (2.4) becomes the one-dimensional wave equation

$$u_{tt} = c_0^2 u_{xx}.$$

By the D'Alembert formula, the general solution of this equation is the sum of two travelling waves

$$u(x, t) = f(x - c_0 t) + g(x + c_0 t),$$

and the dispersion relation yields wave modes with frequencies $\omega_{\pm}(k) = \pm c_0 k$ and phase velocities $c_p = \pm c_0$. In this case, dispersion does not take place. We note that Eq. (2.4) is a second order hyperbolic differential equation written as

$$u_{\xi\eta} = (\alpha^2/4c_0^2)u$$

in the characteristic variables $\xi = x - c_0 t$ and $\eta = x + c_0 t$. In the case $\alpha \neq 0$, dispersion causes distortion of a signal consisting of harmonics with several frequencies. It is less seen for large k (i.e., in the domain of high frequencies ω) since the phase velocities $c_{\pm}(k) = \pm\sqrt{c_0^2 + \alpha^2 k^{-2}}$ are asymptotically constant as $k \to \infty$, i.e., $c_p^{\pm}(k) \to \pm c_0$.

If a frequency defined by the dispersion relation (2.3) is complex, i.e., $\omega = \omega_r + i\omega_i$, then the wave packet has the form

$$u(x, t) = a(t)e^{i(kx - \omega_r t)},$$

where $a(t) = a\exp(\omega_i t)$ is the amplitude factor depending on time. If the imaginary part of the frequency ω is negative ($\omega_i < 0$), then the wave amplitude exponentially decays as $t \to +\infty$, i.e., dissipation takes place.

Example 2.2 Consider the linearized Korteweg-de Vries–Burgers equation

$$u_t + u_0 u_x + u_{xxx} = \nu u_{xx},$$

where u_0 and $\nu > 0$ are constants. A unique wave mode for this equation is given by the frequency

$$\omega(k) = u_0 k - k^3 - i\nu k^2.$$

Since $\mathrm{Im}\,\omega = -\nu k^2 < 0$, the wave packets with $k \neq 0$ exponentially decay, which is caused by the fact that the equation involves the second order derivative u_{xx} with coefficient ν, interpreted as the viscosity of a continuous medium. Long waves (the limit as $k \to 0$) are less susceptible to dissipation, the greater the viscosity, the greater the decay is.

In the case of complex frequencies with $\omega_i > 0$, the amplitude of the solution increases infinitely with time. In this case, instability of the wave process is observed. It should be taken into account that the system (2.1) appearing as a result of linearization of more general nonlinear equations was first designated for describing propagation of small perturbations in a continuous medium. Therefore, a wave packet with $\omega_i > 0$ can simulate the instability process only at the initial stage since the original linear approximation loses sense with growth of perturbations. From this point of view, a neutral-stable elementary wave packet with real frequency ($\omega_i = 0$) and constant amplitude a describes regular propagation of waves in the case where the influence of dissipation on instability is negligible.

2.2   Multi-dimensional Wave Packets

In the case of systems of l differential equations of the form (2.1) for $\mathbf{u} = (u_1, \ldots, u_l)^T$, elementary wave packets are described by the solutions $\mathbf{u}(x, t) = \mathbf{a} \exp i(kx - \omega t)$, where $\mathbf{a} = (a_1, \ldots, a_l)^T$ is the amplitude vector.

Example 2.3 Consider the system of quasilinear first order equations

$$d_t \mathbf{u} + A \partial_x \mathbf{u} + B \partial_x^2 \mathbf{u} = 0 \tag{2.5}$$

for an n-dimensional vector $\mathbf{u} = (v, u_1, u_2, \ldots, u_{n-1})^T$, $n \geqslant 2$, with the differentiation operator $d_t = \partial_t + v \partial_x$. The distinguished component v of the vector-valued function $\mathbf{u}$ is interpreted as the displacement velocity of particles of a continuous medium along the trajectories $dx/dt = v(x, t)$. We assume that constant $n \times n$-matrices A and B are real; moreover, the matrix A is symmetric and the matrix B is antisymmetric: $A = A^T$ and $B = -B^T$. The linearization of the system (2.5) at the rest state $\mathbf{u} = 0$ yields the system of equations with constant coefficients

$$\partial_t \mathbf{u} + A \partial_x \mathbf{u} + B \partial_x^2 \mathbf{u} = 0.$$

We look for a solution in the form of a wave packet $\mathbf{u}(x, t) = \mathbf{a} \exp i(kx - \omega t)$ with real wavenumber k. Then we have the equation

$$(ikB + A - cI)\mathbf{a} = 0$$

for the amplitude vector $\mathbf{a}$, where $c = \omega/k$ is the phase velocity. Solutions with an amplitude $\mathbf{a} \neq 0$ can exist only if the following equality holds:

$$\det(ikB + A - cI) = 0.$$

This equality is the *dispersion relation*. All its roots $c_j = c_j(k)$ $(j = 1, \ldots, n)$ are real in view of properties of the matrices A and B. Consequently, the frequencies

$\omega_j(k) = c_j(k)k$ of all wave modes are also real. We note that the system (2.5) is invariant under the Galilean transform

$$\widetilde{t} = t, \quad \widetilde{x} = x - u_0 t, \quad \widetilde{v} = v - u_0, \quad \widetilde{u}_i = u_i \quad (i = 1, \dots, n - 1),$$

realizing the passage to the coordinate system moving with velocity $u_0 = \text{const}$. The linearization of the system (2.5) at a constant solution $\mathbf{u} = (u_0, 0, \dots, 0)$ yields the system of equations

$$\partial_t \mathbf{u} + \widetilde{A} \partial_x \mathbf{u} + B \partial_x^2 \mathbf{u} = 0$$

with the matrix $\widetilde{A} = A + u_0 I$. The phase velocity $\widetilde{c}$ of wave packets for this system is given by $\widetilde{c} = \widetilde{\omega}/k$, where the wave frequencies $\widetilde{\omega}$ and ω in the moving and fixed coordinate systems are connected by the relation $\widetilde{\omega} = \omega - u_0 k$. In wave theory, the change in frequency is called the *Doppler effect* and the change in frequency $u_0 k$ is referred to as the *Doppler shift*.

In the case $\mathbf{x} \in \mathbb{R}^3$, by an *elementary wave packet* we mean a function of the form

$$u(\mathbf{x}, t) = a e^{i(\mathbf{k} \cdot \mathbf{x} - \omega t)},$$

where $\mathbf{k} = (k_1, k_2, k_3)^T$ is the wave vector. The surface of constant phase is the plane $k_1 x_1 + k_2 x_2 + k_3 x_3 - \omega t = \text{const}$ which is travelling in the space $\mathbb{R}^3$ along the direction of the vector $\mathbf{k}$ with the *normal phase velocity* $c_p = \omega/|\mathbf{k}|$ (cf. Fig. 2.1).

Problem 2.1 Find the normal phase velocities for the three-dimensional wave packets

$$\mathbf{w}(\mathbf{x}, t) = \mathbf{a} \cos(\mathbf{k} \cdot \mathbf{x} - \omega t), \quad \mathbf{a} \in \mathbb{R}^3$$

Fig. 2.1 The wave vector $\mathbf{k}$ is always orthogonal to surfaces of the constant phase

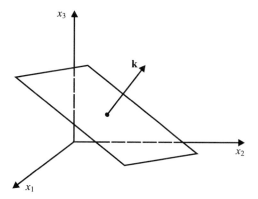

satisfying the system of linear elasticity (the Lamé equations)

$$\rho_0 \mathbf{w}_{tt} = (\lambda + \mu)\nabla \operatorname{div} \mathbf{w} + \mu \triangle \mathbf{w},$$

where $\mathbf{w} = (w_1, w_2, w_3)^T$ is the displacement vector, $\triangle = \partial^2_{x_1} + \partial^2_{x_2} + \partial^2_{x_3}$, $\operatorname{div} \mathbf{w} = w_{1x_1} + w_{2x_2} + w_{3x_3}$, $\nabla = (\partial_{x_1}, \partial_{x_2}, \partial_{x_3})$, $\rho_0 = \operatorname{const}$ is the medium density, $\lambda > 0$ and $\mu > 0$ are the Lamé constants.

Solution We first find the result of action of the main differential operators on a vector-valued function $\mathbf{w}(\mathbf{x}, t)$ in the form of a cosinusoidal wave packet:

$$\triangle \mathbf{w} = -|\mathbf{k}|^2 \mathbf{a} \cos(\mathbf{k} \cdot \mathbf{x} - \omega t),$$
$$\nabla \operatorname{div} \mathbf{w} = -\mathbf{k}(\mathbf{a} \cdot \mathbf{k}) \cos(\mathbf{k} \cdot \mathbf{x} - \omega t).$$

From the Lamé equations it follows that the amplitude vector $\mathbf{a}$, wave vector $\mathbf{k}$, and frequency ω satisfy the identity

$$(\rho_0 \omega^2 - \mu|\mathbf{k}|^2)\mathbf{a} = (\lambda + \mu)\mathbf{k}(\mathbf{a} \cdot \mathbf{k}). \tag{2.6}$$

For a given vector $\mathbf{k} \neq 0$ we represent the amplitude vector in the form

$$\mathbf{a} = |\mathbf{k}|^{-2}(\mathbf{a} \cdot \mathbf{k})\mathbf{k} + \mathbf{b},$$

where the vector $\mathbf{b}$ is orthogonal to the wave vector: $\mathbf{b} \cdot \mathbf{k} = 0$. The projections of the vector identity (2.6) onto the direction of $\mathbf{k}$ and onto the perpendicular plane yield the system of equations

$$(\rho_0 \omega^2 - (\lambda + 2\mu)|\mathbf{k}|^2)\mathbf{a} \cdot \mathbf{k} = 0,$$
$$(\rho_0 \omega^2 - \mu|\mathbf{k}|^2)\mathbf{b} = 0.$$

If $\mathbf{a} \cdot \mathbf{k} \neq 0$, then $\omega^2 = (\lambda + 2\mu)|\mathbf{k}|^2/\rho_0$ and, consequently, $\mathbf{b} = 0$. In this case, the direction of the displacement vector $\mathbf{w}$ coincides with the direction of $\mathbf{k}$, i.e., the wave packet is a longitudinal wave propagating with the normal phase velocity

$$c_p = \pm\sqrt{(\lambda + 2\mu)/\rho_0}.$$

If $\mathbf{a} \cdot \mathbf{k} = 0$, but $\mathbf{a} \neq 0$, then $\mathbf{b} \neq 0$. Therefore, $\omega^2 = \mu|\mathbf{k}|^2/\rho_0$. In this case, we have a transverse wave such that the displacement vector $\mathbf{w}$ is orthogonal to the direction of the wave propagation, whereas the normal phase velocity is equal to $c_p = \pm\sqrt{\mu/\rho_0}$. ◻

Answer:

$$c_p^2 = (\lambda + 2\mu)/\rho_0 \quad (\mathbf{a} \cdot \mathbf{k} \neq 0)$$
$$c_p^2 = \mu/\rho_0 \quad (\mathbf{a} \cdot \mathbf{k} = 0).$$

2.3   Group Velocity

The dispersion property is highlighted in the interaction of the wave packets $u(x,t) = a\cos(kx - \omega t)$ of a fixed mode with real frequency $\omega = \omega(k)$ and the same amplitude, but different wavenumbers k. For the sum of such two packets we have

$$a\cos(kx - \omega t) + a\cos(k_1 x - \omega_1 t)$$

$$= 2a\cos\left(\frac{k_1 - k}{2}x - \frac{\omega_1 - \omega}{2}t\right)\cos\left(\frac{k_1 + k}{2}x - \frac{\omega_1 + \omega}{2}t\right),$$

where $\omega_1 = \omega(k_1)$. The wave motion described by this sum is represented as a periodic sequence of groups of waves propagating with the velocity $(\omega_1 - \omega)/(k_1 - k)$. As $k_1 \to k$, the velocity of the envelope coincides with the derivative

$$c_g(k) = \frac{d\omega}{dk},$$

called the *group velocity*. The maximal amplitude of crests of the carrying wave for each group is approximately equal to the double amplitude of the original wave packets, whereas their displacement velocity is equal to the phase velocity $c_p(k)$ (cf. Fig. 2.2). The group and phase velocities are connected by the relation

$$c_g = c_p + k\frac{dc_p}{dk}.$$

Hence for dispersive waves the group and phase velocities are different.

Problem 2.2 Consider the linear Boussinesq equation describing long shallow water waves of small amplitude

$$u_{tt} - c_0^2 u_{xx} = \frac{1}{3}h_0^2 u_{xxtt},$$

Fig. 2.2 For dispersive
waves the group and phase
velocities are different

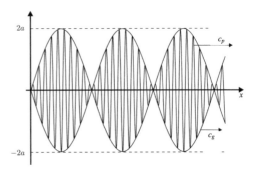

where h_0 is the depth of the fluid at rest, g is the acceleration of gravity, and $c_0 = \sqrt{gh_0}$. Find the phase velocity c_p and the group velocity c_g. What are the maximal values $|c_p|$ and $|c_g|$?

Solution Looking for a solution in the form of an elementary wave packet

$$u(x, t) = a\exp(i(kx - \omega t)),$$

we obtain the dispersion relation

$$\omega^2 \left(1 + \frac{1}{3}h_0^2 k^2 \right) = c_0^2 k^2$$

which determines two modes corresponding to the waves travelling to the left and to the right respectively. Hence for each mode we can find the phase velocity

$$c_p(k) = \pm c_0 \left(1 + \frac{1}{3}h_0^2 k^2 \right)^{-1/2}$$

and the group velocity

$$c_g(k) = \pm c_0 \left(1 + \frac{1}{3}h_0^2 k^2 \right)^{-3/2}.$$

Since the velocities do not coincide for an arbitrary $k \neq 0$, the waves are dispersive. Furthermore, the following estimates hold:

$$|c_p(k)| \leq c_0, \quad |c_g(k)| \leq c_0,$$

where equality is attained at the long-wave limit $k = 0$.

   Thus, the absolute values of the phase and group velocities do not exceed the critical velocity $\sqrt{gh_0}$.   □

In the multi-dimensional case, the group velocity is defined by the equality

$$\mathbf{c}_g(\mathbf{k}) = \nabla \omega(\mathbf{k}).$$

In the case of spatial dispersive waves, not only absolute values, but also directions of the phase velocity vector $\mathbf{c}_p(\mathbf{k}) = (\omega(\mathbf{k})/|\mathbf{k}|)\mathbf{k}$ and the group velocity vector $\mathbf{c}_g(\mathbf{k})$ can be different.

Example 2.4 Let a frequency ω be a homogeneous function of degree 0 in $\mathbf{k} = (k, l, m)^T$, i.e.,

$$\omega(\lambda k, \lambda l, \lambda m) = \omega(k, l, m)$$

for all $\lambda > 0$. Differentiating this identity with respect to the parameter λ at the point $\lambda = 1$, we obtain the relation

$$k\frac{\partial \omega}{\partial k} + l\frac{\partial \omega}{\partial l} + m\frac{\partial \omega}{\partial m} = 0 \tag{2.7}$$

which means that the vectors $\mathbf{c}_g(\mathbf{k})$ and $\mathbf{k}$ are perpendicular at each point $\mathbf{k} \in \mathbb{R}^3$. The homogeneity of the frequency is also a necessary condition for the orthogonality of the group velocity and the wave vector. Indeed, (2.7) is a linear first order partial differential equation for $\omega(k, l, m)$. The equations of characteristics have the form $dk/k = dl/l = dm/m$, which implies that the general solution $\omega = \omega(k/l, k/m)$ is a homogeneous function of degree 0 in the variables k, l, m.

2.4   Stationary Phase Method

A wave process in a dispersive medium usually becomes regular with time even if the wave propagation is caused by an initial perturbation of a general form. For a given mode $\omega = \omega(k)$ we consider the linear superposition of elementary wave packets

$$u(x, t) = \int_{-\infty}^{+\infty} a(k)e^{i(kx - \omega(k)t)} \, dk. \tag{2.8}$$

The amplitude factor $a(k)$ is uniquely determined by the initial function $u(x, 0)$ via the Fourier integral

$$a(k) = \frac{1}{2\pi} \int_{-\infty}^{+\infty} u(x, 0)e^{-ikx} \, dx.$$

It is convenient to study the behavior of the solution (2.8) for large t for a fixed ratio $x/t = U$, which corresponds to the motion of the observer with constant velocity U. We introduce the phase function $\psi(k) = kx/t - \omega(k)$. Then for $v(t) = u(Ut, t)$ we have

$$v(t) = \int_{-\infty}^{+\infty} a(k)e^{it\psi(k)}\, dk.$$

We assume that $a(z)$ and $\psi(z)$ are analytic functions of the complex variable $z = k + il$ in the strip $|\operatorname{Im} z| < l_0$, where $l_0 > 0$. One can show that the contribution of the integration intervals where $\psi'(k) \neq 0$ to the solution is exponentially small as $t \to +\infty$. In a neighborhood of a point k_0 where $\psi'(k_0) = 0$, but $\psi''(k_0) \neq 0$, we have the expansion

$$\psi(z) = \psi(k_0) + \frac{1}{2}(z - k_0)^2\psi''(k_0) + O(|z - k_0|^3).$$

Therefore, the structure of level lines $\operatorname{Im}\psi(z) = C$ in a neighborhood of such a stationary point is similar to the structure of the level lines of the saddle surface $(k - k_0)l = C/\psi''(k_0)$. The level lines of the function $\operatorname{Im}\psi$ in a neighborhood of the point $(k_0, 0)$, where $\psi''(k_0) < 0$, are shown in Fig. 2.3.

For solutions of the form (2.8) the phase of the wave packet under the integral sign is stationary at the point k_0 where $\omega'(k_0) = U = x/t$, which corresponds to the point $x = x(t)$ moving with the group velocity $c_g(k_0)$. Deforming the integration contour along the real axis in (2.8) to a contour in the complex domain, as shown in Fig. 2.3, and computing the integral corresponding to the first two nonzero terms in

Fig. 2.3 The integration contour along the real axis in (k, l)-plane is deformed into the other passing through the saddle point $(k_0, 0)$ (*bold solid lines*)

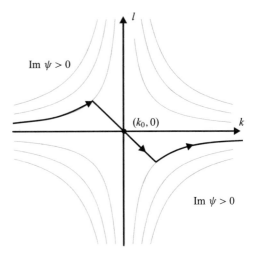

the Taylor expansion of the phase function $\psi(z)$, we obtain the asymptotic expansion

$$u(x,t) = \sqrt{\frac{2\pi}{|\omega''(k_0)|t}} a(k_0) e^{i\left(k_0 x - \omega(k_0)t - \frac{\pi}{4}\,\mathrm{sgn}\,\omega''(k_0)\right)} + O\left(\frac{1}{t}\right). \tag{2.9}$$

Consequently, the initial perturbation is divided into groups of wave packets with an amplitude decreasing like $1/\sqrt{t}$ as $t \to +\infty$. In the case of dispersive media, the carried energy density is, as a rule, proportional to the squared amplitude of the solution $|u(x,t)|^2$. By the above asymptotics of the solution, the part of energy contained between the straight lines $x_j = \omega'(k_j)t$ $(j = 1, 2)$, is preserved in the first approximation:

$$\int_{x_1}^{x_2} |u(x,t)|^2 dx = 2\pi \int_{k_1}^{k_2} |a(k)|^2 dk.$$

This means that the energy is carried with the group velocity; moreover, dispersion of a wave implies scattering of the energy in the space.

Problem 2.3 Consider the Cauchy problem for the Euler–Bernoulli beam equation

$$u_{tt} + \gamma^2 u_{xxxx} = 0 \quad (-\infty < x < +\infty),$$
$$u(x,0) = u_0(x), \quad u_t(x,0) = 0,$$

where the constant $\gamma > 0$ characterizes elastic properties of the material and $u_0(x) \in C^\infty$ is an absolutely integrable and even function of x. Using the stationary phase method, find an asymptotics of the solution as $t \to +\infty$.

Solution The direct and inverse Fourier transforms of a function $u(x,t)$ with respect to x have the form

$$\widehat{u}(k,t) = \frac{1}{2\pi} \int_{-\infty}^{+\infty} u(x,t) e^{-ikx}\, dx,$$

$$u(x,t) = \int_{-\infty}^{+\infty} \widehat{u}(k,t) e^{ikx}\, dk.$$

Since $\widehat{\partial_x^n u}(k,t) = (ik)^n \widehat{u}(k,t)$, the Fourier-image $\widehat{u}(k,t)$ is a solution to the Cauchy problem for the second order ordinary differential equation in t:

$$\widehat{u}_{tt} + \gamma^2 k^4 \widehat{u} = 0,$$
$$\widehat{u}(k,0) = \widehat{u}_0(k), \quad \widehat{u}_t(k,0) = 0.$$

Integrating, we find

$$\widehat{u}(k,t) = \widehat{u}_0(k) \cos \gamma k^2 t = \frac{1}{2} \widehat{u}_0(k)\{e^{i\gamma k^2 t} + e^{-i\gamma k^2 t}\}.$$

Applying the inverse Fourier transform, we can represent the solution as

$$u(x,t) = \frac{1}{2}\{I_+(x,t) + I_-(x,t)\},$$

where

$$I_\pm(x,t) = \int\limits_{-\infty}^{+\infty} \widehat{u}_0(k) e^{i(kx - \omega_\pm(k)t)} dk, \quad \omega_\pm(k) = \pm \gamma k^2.$$

Each integral $I_\pm$ can be represented in the form (2.8), and it is possible to use the asymptotic formula (2.9) with the wavenumber $k_0 = k_0(x,t)$ defined implicitly by the relation $x/t = \omega'(k_0)$. For the wave modes $\omega_\pm(k) = \pm \gamma k^2$ this relation implies $k_0(x,t) = \pm x/(2\gamma t)$. Furthermore, for these modes $\omega''_\pm(k_0) = \pm 2\gamma \neq 0$. Therefore, by formula (2.9),

$$I_\pm(x,t) = \sqrt{\frac{\pi}{\gamma t}} \widehat{u}_0\left(\pm \frac{x}{2\gamma t}\right) e^{\pm i\left(\frac{x^2}{4\gamma t} - \frac{\pi}{4}\right)} + O\left(\frac{1}{t}\right) \quad (t \to +\infty).$$

It remains to note that if the initial function $u_0(x)$ is even in x, then its Fourier transform is even in k:

$$a(k) \stackrel{\text{def}}{=} \widehat{u}_0(k) = \frac{1}{2\pi} \int\limits_{-\infty}^{+\infty} u_0(x) \cos kx\, dx = a(-k).$$

Taking into account this evenness property, we obtain the desired asymptotic expansion of the solution.   □

Answer:

$$u(x,t) = \sqrt{\frac{\pi}{\gamma t}} a\left(\frac{x}{2\gamma t}\right) \cos\left(\frac{x^2}{4\gamma t} - \frac{\pi}{4}\right) + O\left(\frac{1}{t}\right),$$

where

$$a(k) = \frac{1}{\pi} \int\limits_{0}^{+\infty} u_0(x) \cos kx\, dx.$$

The solution behaves according to formula (2.9) as $t \to +\infty$ if the condition $\omega''(k_0) \neq 0$ holds. This condition fails at extremum points of the group velocity, where $\omega''(k_0) = c_g'(k_0) = 0$. In neighborhoods of such stationary points, the solution has another asymptotics. To construct the asymptotics, we introduce the phase function ψ in some other way, namely: $\psi(k) = k\omega'(k_0) - \omega(k)$ so that the wave phase $\theta = kx - \omega t$ is connected with ψ by the equality

$$\theta = k(x - c_g(k_0)t) + \psi(k)$$

(it is assumed that x/t is constant). Since $\psi'(k_0) = 0$ and $\psi''(k_0) = 0$, we have the following decomposition of $\psi(z)$:

$$\psi(z) = \psi(k_0) + \frac{1}{6}(z - k_0)^3 \psi'''(k_0) + O(|z - k_0|^4).$$

Analyzing the structure of level lines of the function Im ψ and taking into account the above decomposition and suitable deformation of the integration contour in (2.8) in the complex plane $z = k + il$, we can characterize the behavior of the solution $u(x, t)$ as $t \to +\infty$. Namely, in a neighborhood of the point $x = c_g(k_0)t$ such that $c_g'(k_0) = 0$, $c_g''(k_0) \neq 0$, we have

$$u(x, t) = \frac{2\pi a(k_0)}{\sqrt[3]{(1/2)|c_g''(k_0)|t}} Ai\left(\frac{x - c_g(k_0)t}{\sqrt[3]{(1/2)|c_g''(k_0)|t}}\right) e^{i(k_0 x - \omega(k_0)t)} + O(t^{-\frac{2}{3}}). \qquad (2.10)$$

This formula involves the Airy function

$$Ai(x) = \frac{1}{\pi} \int\limits_0^{+\infty} \cos\left(kx + \frac{1}{3}k^3\right) dk$$

with the following behavior as $|x| \to \infty$:

$$Ai(x) \sim \frac{1}{2\sqrt{\pi}|x|^{\frac{1}{4}}} \begin{cases} \exp\left(-\frac{2}{3}x^{\frac{3}{2}}\right), & x \to +\infty, \\ 2\cos\left(\frac{2}{3}|x|^{\frac{3}{2}} - \frac{\pi}{4}\right), & x \to -\infty. \end{cases}$$

The graph of the Airy function (cf. Fig. 2.4) has the shape of the envelope of wave packets travelling with group velocity $c_g(k_0)$. Hence it is clear that the front of the first-localized perturbation propagates with the velocity equal to the extremal value of the group velocity; moreover, the distance between neighboring zeros of the envelope in the domain of its oscillations unboundedly increases with the growth of t. By the approximate formula (2.10) and the above asymptotic formula for the Airy function, the wave packets decay in a neighborhood of the front (like $t^{-1/3}$) slightly

Fig. 2.4 The graph of the
Airy function is shown. It has
the shape of the envelope of
wave packets travelling with
group velocity

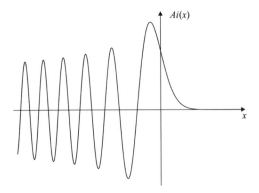

slower than inside the domain of the wave motion (like $t^{-1/2}$). In the unperturbed
domain before the front, the wave process exponentially decays with increasing
distance from the front.

2.5 Nonlinear Dispersion

Decomposition of monochromatic wave packets into groups of waves caused by
their interaction is typical for the evolution of a wave field in a medium with
dispersion. Moreover, the behavior of the envelope of wave packets is often
described by wave type differential equations. Thus, to describe modulated wave
packets of the form

$$u(x,t) = \psi(x,t) \exp i(k_0 x - \omega(k_0)t)$$

with a complex-valued amplitude ψ smoothly depending on x and t, many models of
continuum mechanics and mathematical physics involve the nonlinear Schrödinger
equation

$$i\psi_t + \psi_{xx} \pm |\psi|^2 \psi = 0. \tag{2.11}$$

The equations in (2.11) with the sign "minus" or "plus" at the nonlinear term have
different mathematical properties. The equation itself has solutions in the form of
wave packets $\psi(x,t) = a \exp i(kx - \omega t)$ with a constant amplitude a independent
of x and t. Substituting ψ into (2.11), we obtain the dispersion relation

$$D(\omega, k; a) = \omega - k^2 \pm a^2 = 0$$

which, in addition to the frequency ω and wavenumber k, contains the wave
amplitude a. This fact expresses one of the main differences between the case of
nonlinear dispersive waves and the linear case.

Problem 2.4 For the nonlinear Schrödinger equation (2.11) with the sign "plus" construct an envelope solitary wave $\psi(x, t) = A(x - ct) \exp i(kx - \omega t)$, where A is a real amplitude function A such that $A, A' \to 0$ as $|x| \to \infty$.

Solution Substituting the function ψ into (2.11), we obtain the complex ordinary second order differential equation

$$A'' - k^2 A + \omega A + A^3 + i(2k - c)A' = 0.$$

Since A is real, the imaginary part of this equation yields the relation $(2k - c)A' = 0$. Consequently, for the existence of a nonconstant solution $A \neq \text{const}$ it is necessary that $c = 2k$ for the velocity c of the envelope solitary wave. Taking into account this fact, we obtain the nonlinear equation

$$A'' - \beta^2 A + A^3 = 0,$$

where $\beta^2 = k^2 - \omega > 0$. Multiplying the equation for A by A', integrating, and taking into account the decay at infinity, we obtain a first order equation with separated variables. The solution of this equation is given by the quadrature formula

$$\int \frac{dA}{A\sqrt{\beta^2 - (1/2)A^2}} = \pm\xi \quad (\xi = x - ct), \tag{2.12}$$

where the integral is computed by substitution $A = \beta/(\sqrt{2}\cosh\beta\xi)$. Thus, we find the required envelope $A(\xi) = a/\cosh(\sqrt{2}a\xi)$, where the amplitude $a = \beta/(\sqrt{2})$ is connected with the frequency ω and wavenumber k of the carrying wave packet by the dispersion relation $\omega = k^2 - 2a^2$. Moreover, the displacement velocity of the envelope coincides with the group velocity $c_g(k) = 2k$ of the linearized Schrödinger equation $i\psi_t + \psi_{xx} = 0$. ☐

Answer:

$$\psi(x, t) = \pm\frac{ae^{i(kx - \omega t)}}{\cosh\{\sqrt{2a}(x - ct)\}},$$

where $\omega = k^2 - 2a^2$, $c = 2k$.

To describe nonlinear long waves with dispersion, one often uses the Korteweg-de Vries equation

$$u_t + uu_x + u_{xxx} = 0. \tag{2.13}$$

The joint role of nonlinearity and dispersion is already apparent while searching a travelling wave type solution. Substituting $u = u(x - ct)$, $c > 0$, into (2.13), we obtain the third order ordinary differential equation

$$-cu' + uu' + u''' = 0.$$

The double integration reduces this equation to the first order equation

$$u'^2 = \frac{1}{3}(u - u_1)(u - u_2)(u_3 - u), \tag{2.14}$$

where the roots of the cubic polynomial on the right-hand side are connected with the integration constants and the wave velocity c by the Vieta formula (in particular, $u_1 + u_2 + u_3 = 3c$). Solutions in the form of a periodic wave are obtained for simple real roots $u_1 < u_2 < u_3$. Without loss of generality we can assume that $u_2 = 0$ since Eq. (2.13) is invariant under the Galilean transform $\widetilde{t} = t, \widetilde{x} = x - u_2 t, \widetilde{u} = u - u_2$. Denoting by $a = u_3$ a positive root $u_3 > 0$, we have $u_1 = 3c - a$ for the negative root $u_1 < 0$. Under this choice of the reference system in the travelling wave, we can reduce Eq. (2.14) to the quadrature formula

$$\pm\sqrt{3} \int\limits_u^a \frac{ds}{\sqrt{(s + a - 3c)s(a - s)}} = x - ct.$$

In the general case, this integral is not necessarily expressed in terms of elementary functions. However, substituting $s = a\cos^2\theta$ and making similar change of the sought function $u = a\cos^2\varphi$, we can reduce this dependence to the relation

$$\int\limits_0^\varphi \frac{d\theta}{\sqrt{1 - m^2\sin^2\theta}} = \kappa\xi \quad (\xi = x - ct), \tag{2.15}$$

where

$$m^2 = \frac{u_3 - u_2}{u_3 - u_1} = \frac{a}{2a - 3c}, \quad \kappa^2 = \frac{2a - 3c}{12}.$$

The function $\varphi = am(\kappa\xi; m)$, defined implicitly by (2.15), is called the *Jacobi amplitude*, and the superposition $cn(\kappa\xi; m) = \cos am(\kappa\xi; m)$ is called the *elliptic cosine*. Consequently, the required wave profile has the form

$$u(x, t) = acn^2(\kappa(x - ct); m).$$

Such a periodic wave is said to be *cnoidal* because its definition involves the function cn. For a given value of m $(0 < m < 1)$ the function $cn(\cdot; m)$ is periodic with period $2K(m)$, where

$$K(m) = \int\limits_0^{\frac{\pi}{2}} \frac{d\theta}{\sqrt{1 - m^2 \sin^2 \theta}}$$

is a complete elliptic integral. Hence the parameter $a > 0$ characterizing the wave height, i.e., the difference between the elevation of a crest and a neighboring trough (the double amplitude) is connected with the wave velocity c and period L with respect to the independent variable x by the relation

$$L = \frac{4\sqrt{3}}{\sqrt{2a - 3c}} K\left(\frac{\sqrt{a}}{\sqrt{2a - 3c}}\right) \quad (3c < a). \tag{2.16}$$

If we formally introduce the wavenumber $k = 2\pi/L$, then this equality can be regarded as the dispersion relation for cnoidal waves. Elliptic Jacobi functions are reduced to elementary functions as $m \to 0$ and $m \to 1$ (the parameter m is called the *modulus of an elliptic function*). The function $cn(\zeta; m)$ uniformly converges to the function $\cos \zeta$ as $m \to 0$, which corresponds to transformation of a cnoidal wave of small amplitude $a \to 0$ to a usual wave packet in the linear theory. On each finite interval of the variable ζ, the function $cn(\zeta; m)$ uniformly converges to $1/\cosh \zeta$ as $m \to 1$. In this case, the periodic wave degenerates into the travelling solitary wave

$$u(x, t) = 3c \cosh^{-2}\{(\sqrt{c}/2)(x - ct)\}.$$

A generalization of solitary wave type solutions are exact multi-soliton solutions to the Korteweg-de Vries equation describing nonstationary interaction between solitary waves. In soliton theory, it is convenient to write the Korteweg-de Vries equation as

$$u_t + 6uu_x + u_{xxx} = 0. \tag{2.17}$$

n-Soliton solutions to this equation have the form

$$u(x, t) = 2\frac{\partial^2}{\partial x^2} \ln \det \{I + A(x, t)\},$$

where I is the identity matrix of order n and A is a symmetric matrix with entries

$$a_{ij}(x, t) = \frac{\gamma_i \gamma_j}{k_i + k_j} e^{-(k_i + k_j)x} e^{4(k_i^3 + k_j^3)t} \quad (i, j = 1, \ldots, n),$$

where $\gamma_1, \ldots, \gamma_n$ are arbitrary real constants and $k_1, \ldots, k_n$ $(k_i + k_j \neq 0)$.

2.6   Problems

1. Find the phase velocity $c_p(k) = \omega(k)/k$ and the group velocity $c_g(k) = d\omega/dk$
    for the dispersion relations $\omega = \omega(k)$ generated by the following equations:

    (a) $u_t + c_0 u_x + c_0 h_0^2 u_{xxx} = 0$,
    (b) $u_t + c_0 u_x - h_0^2 u_{xxt} = 0$,

    where $c_0 > 0$ and $h_0 > 0$ are constants. Show that for all $k \geq 0$

    $$\omega^{(a)}(k) \leq \omega^{(b)}(k),$$

    $$c_p^{(a)}(k) \leq c_p^{(b)}(k),$$

    $$c_g^{(a)}(k) \leq c_g^{(b)}(k),$$

    where equalities hold only for $k = 0$. What are the limits of the ratios

    $$\frac{c_g^{(a)}(k)}{c_p^{(a)}(k)}, \qquad \frac{c_g^{(b)}(k)}{c_p^{(b)}(k)}$$

    in the long-wave ($k \to 0$) and short-wave ($k \to \infty$) approximations? Construct
    the graphs of the functions $\omega(k)$, $c_p(k)$, and $c_g(k)$ for $-\infty < k < +\infty$ in cases
    (a) and (b).
    Answer:

    $$\lim_{k \to 0} \frac{c_g^{(a)}(k)}{c_p^{(a)}(k)} = \lim_{k \to 0} \frac{c_g^{(b)}(k)}{c_p^{(b)}(k)} = 1,$$

    $$\lim_{k \to \infty} \frac{c_g^{(a)}(k)}{c_p^{(a)}(k)} = 3, \qquad \lim_{k \to \infty} \frac{c_g^{(b)}(k)}{c_p^{(b)}(k)} = -1.$$

2. It is known that the group velocity of waves described by a differential equation
    with two independent variables x, t and complex constant coefficients is real
    and coincides with the double phase velocity. Restore the form of the equation
    provided that it has (a) the first order in t, (b) the second order in t.
    Answer:

    (a) $iu_t + \gamma u_{xx} = 0$,
    (b) $u_{tt} - i(\gamma_1 + \gamma_2)u_{xxt} - \gamma_1\gamma_2 u_{xxxx} = 0$

    where $\gamma, \gamma_1, \gamma_2 \in \mathbb{R}$ are constants.
3. Show that for the equation

    $$u_{tt} - c_0^2 u_{xx} + u = 0,$$

where $x_1, x_2 \in \mathbb{R}$, the energy balance equation holds:

$$\frac{d}{dt} \int_{x_1}^{x_2} e(x,t)dx + f(x,t) \Big|_{x_1}^{x_2} = 0,$$

where $e = (u_t^2 + c_0^2 u_x^2 + u^2)/2$ is the energy density and $f = -c_0^2 u_x u_t$ is the energy flux.

4. The velocity of energy transfer by the wave packet $u = a\sin(kx - \omega t)$ is equal to the ratio $U = F/E$ of the mean energy flux over the time period $T = 2\pi/\omega$ to the mean energy density E over the space period $L = 2\pi/k$,

$$F = \frac{1}{T} \int_{t_1}^{t_1+T} f(x,t)dt, \quad E = \frac{1}{L} \int_{x_1}^{x_1+L} e(x,t)dx.$$

Under the conditions of the previous problem, show that U coincides with the group velocity $c_g(k)$.

5. Show that the local frequency ω, local wave number k, and phase θ of the group of waves defined implicitly as functions of x and t by the relations

$$\omega = W(k), \quad x = W'(k)t, \quad \theta = kx - \omega t,$$

satisfy the differential equations

$$k_t + W'(k)k_x = 0,$$
$$\theta_x = k,$$
$$\theta_t = -\omega$$

provided that $W''(k) \neq 0$.

6. It is known that for a mode $\omega = \omega(\mathbf{k})$ with two-dimensional wave vector $\mathbf{k} = (k,l)^T$ the group velocity vector $\mathbf{c}_g = (\partial\omega/\partial k, \partial\omega/\partial l)^T$ for all $\mathbf{k} \neq 0$ forms a constant angle α ($0 \leqslant \alpha \leqslant \pi$) with the wave vector. For what dependence $\omega(\mathbf{k})$ is it possible?

   Answer: $\omega(\mathbf{k}) = f\left(|\mathbf{k}|e^{\pm\varphi\tan\alpha}\right)$, where $\varphi = \arctan(k/l)$ and the function $f \in C^1$ satisfies the conditions $f'(\xi) \geqslant 0$ ($\xi \in \mathbb{R}$) for $0 \leqslant \alpha < \pi/2$, $f'(\xi) \leqslant 0$ ($\xi \in \mathbb{R}$) for $\pi/2 < \alpha \leqslant \pi$, and $\omega(\mathbf{k}) = f(k/l)$ for $\alpha = \pi/2$.

7. Show that for the dispersion curve $\omega = \omega(k)$ that satisfies the condition $\omega(0) = 0$ and is convex for $k > 0$ the group velocity $c_g(k) = \omega'(k)$ and the phase velocity $c_p(k) = \omega(k)/k$ possess the following properties:

   a. $c_g(k) > c_p(k)$ if $\omega''(k) > 0$ for all $k > 0$ (the dispersion curve is convex).
   b. $c_g(k) < c_p(k)$ if $\omega''(k) < 0$ for all $k > 0$ (the dispersion curve is concave).

8. Show that the resonance of the phase and group velocities $c_p(k_0) = c_g(k_0)$ at a point $k_0 \neq 0$ is equivalent to any of the following properties.

   a. At the point $k = k_0$, the dispersion curve $\omega = \omega(k)$ touches the line passing through the origin in the (k, ω)-plane.
   b. The value $k = k_0$ of the wave number is a stationary point of the phase velocity, i.e., $c_p'(k_0) = 0$.

9. For a real mode $\omega = \omega(k)$ consider the wave packets

$$u_j(x, t) = a_j e^{i(k_j x - \omega(k_j)t)} \quad (k_j \in \mathbb{R}; \ a_j \in \mathbb{C}, \ a_j \neq 0; \ j = 1, 2, 3).$$

   Let m and n be given integers. Find the wavenumbers k_j for which there is a constant $\lambda \in \mathbb{C}$ such that $(u_1(x, t))^m \cdot (u_2(x, t))^n = \lambda u_3(x, t)$ identically with respect to $x, t \in \mathbb{R}$.
   Answer: $k_3 = mk_1 + nk_2$, $\omega(k_3) = m\omega(k_1) + n\omega(k_2)$.

10. Show that for isotropic waves with the dependence $\omega = \omega(|\mathbf{k}|)$, $\omega(0) = 0$, if $\omega'(k) > 0$, $\omega''(k) > 0$, then there exist vectors $\mathbf{k}_1, \mathbf{k}_2, \mathbf{k}_3$ satisfying the three-wave resonance condition

$$\mathbf{k}_3 = \mathbf{k}_1 + \mathbf{k}_2, \quad \omega(|\mathbf{k}_3|) = \omega(|\mathbf{k}_1|) + \omega(|\mathbf{k}_2|),$$

   but this condition cannot be satisfied if $\omega'(k) > 0$, $\omega''(k) < 0$.

11. For what values of real parameters a and b are all wave modes of the following system stable

$$u_{tt} - a^2 u_{xx} + v_{xx} = 0,$$
$$v_{tt} - b^2 v_{xx} + u_{xx} = 0?$$

   Answer: $a^2 b^2 \geqslant 1$.

12. For the system of equations

$$\mathbf{u}_t + A\mathbf{u}_x = C\mathbf{u}$$

   with symmetric real constant $n \times n$-matrices A and C we look for a solution in the form of a wave packet $\mathbf{u}(x, t) = \mathbf{U} \exp ik(x - ct)$ with a wavenumber $k \in \mathbb{R}$ ($k \neq 0$) and an amplitude vector $\mathbf{U} \in \mathbb{C}^n$. Show that the imaginary part c_{im} of the complex phase velocity $c = c_{re} + ic_{im}$ is equal to

$$c_{im} = \frac{(C\mathbf{U}, \mathbf{U})}{k\|\mathbf{U}\|^2},$$

where

$$(\mathbf{U}, \mathbf{V}) = \sum_{j=1}^{n} U_j \overline{V}_j$$

is the inner product of vectors $\mathbf{U}, \mathbf{V} \in \mathbb{C}^n$ (the bar means complex conjugation) and $\|\mathbf{U}\|^2 = (\mathbf{U}, \mathbf{U})$. What is c_{im} if the matrix A is symmetric and the matrix C is antisymmetric?

13. Consider the system of equations

$$A\mathbf{u}_t + B\mathbf{u}_x + C\mathbf{u}_{xx} = 0$$

with symmetric real constant $n \times n$-matrices A and B (the matrix A is positive definite) and antisymmetric matrix C. Show that all wave modes of this system are stable.

14. Consider the wave packet $u = a \exp i(kx - \omega t)$ for the equation

$$u_t - \nu u_{xx} + \sum_{s=0}^{n} b_s \partial_x^{2s+1} u = 0,$$

where $\nu > 0$ and b_s $(s = 1, \ldots, n)$ are constants. For what time T is the absolute value $|u(x, t)|$ halved?
Answer: $T = \ln 2/(\nu k^2)$.

15. For the equation

$$w_{tt} - c_0^2 w_{xx} + c_0^2 \tau w_{xxt} = 0$$

describing waves in a Kelvin–Voigt viscoelastic medium (w is the displacement, c_0 is the velocity of the propagation of longitudinal waves, and $\tau > 0$ is the relaxation parameter) we look for a solution in the form of the wave packet

$$w(x, t) = ae^{i(\varkappa x - \omega t)} = ae^{-\alpha x} e^{i(kx - \omega t)}$$

decaying with the growth of x, where $\varkappa = k + i\alpha$ is the complex wavenumber. Derive the dispersion relation for $\varkappa$ and a real frequency ω. Find the phase velocity of the wave $c_p = \omega/k$ and the decay exponent α regarded as a function of frequency ω.
Answer:

$$\omega^2 = c_0^2(1 + i\tau\omega)\varkappa^2, \quad c_p^2 = \frac{2c_0^2(1 + \tau^2\omega^2)}{1 + \sqrt{1 + \tau^2\omega^2}},$$

$$\alpha^2 = \frac{\omega^2}{2c_0^2} \left(\frac{1}{\sqrt{1 + \tau^2\omega^2}} - \frac{1}{1 + \tau^2\omega^2} \right).$$

16. Perform a similar analysis for the equation

$$\sigma_{tt} - c_0^2 \sigma_{xx} + \frac{1}{\tau} \sigma_t = 0$$

describing waves in a Maxwell viscoelastic medium (here, σ is the stress in the medium).

Answer:

$$\omega^2 + \frac{i\omega}{\tau} = c_0^2 \varkappa^2, \quad c_p^2 = \frac{2c_0^2 \tau \omega}{\sqrt{1 + \tau^2 \omega^2} + \tau \omega},$$

$$\alpha^2 = \frac{\omega}{2c_0^2 \tau} \frac{1}{\sqrt{1 + \tau^2 \omega^2} + \tau \omega}.$$

17. The initial wave profile u_0 has the form of a sequence of triangular impulses defined on pairwise disjoint intervals $I_m = (x_m - 2, x_m + 2)$ $(m = 1, \ldots, n)$:

$$u_0(x) = \begin{cases} 2 - |x - x_m|, & x \in I_m, \\ 0, & x \notin I_m \ (m = 1, \ldots, n). \end{cases}$$

Find the amplitude factor $a(k)$ for the function u_0.

Answer:

$$a(k) = \frac{2 \sin^2 k}{\pi k^2} \sum_{m=1}^{n} e^{-ikx_m}.$$

18. Find the amplitude factor $a(k)$ if the initial function u_0 has the form of a modulated wave packet:

a. $u_0(x) = e^{-|x|} \cos k_0 x$,
b. $u_0(x) = e^{-x^2/2} \cos k_0 x$ $(k_0 = \text{const})$.

Answer:

a. $a(k) = \dfrac{1 + k^2 + k_0^2}{\pi[1 + (k + k_0)^2][1 + (k - k_0)^2]}$,

b. $a(k) = \dfrac{1}{\sqrt{2\pi}} e^{-\frac{1}{2}(k^2 + k_0^2)} \cosh(kk_0)$.

19. Prove that for the initial function $u_0(x)$ such that

$$A \overset{\text{def}}{=} \int_{-\infty}^{+\infty} e^{l_0|x|} |u_0(x)| \, dx < \infty \quad (l_0 > 0)$$

the amplitude function $a(k)$ admits an analytic extension to the strip $|\text{Im } z| < l_0$ of the complex variable $z = k + il$, where the following estimates hold:

$$|a(z)| \leq \frac{A}{2\pi},$$

$$|a'(z)| \leq \frac{A}{2\pi e(l_0 - |\text{Im } z|)}.$$

20. Let the amplitude function a and phase function ψ satisfy the conditions

$$a(k) \in C^1, \quad \psi(k) \in C^2, \quad \psi'(k) \neq 0 \quad (-\infty < k_1 \leq k \leq k_2 < +\infty).$$

Show that

$$\int_{k_1}^{k_2} a(k)e^{it\psi(k)}\,dk = \left.\frac{a(k)e^{it\psi(k)}}{it\psi'(k)}\right|_{k=k_1}^{k=k_2} + o\left(\frac{1}{t}\right) \quad \text{as } t \to +\infty.$$

Hint. Use the Riemann–Lebesgue lemma.

21. The wave motion is described by a solution $u(x, t)$ of the form (2.8) with mode $\omega = \omega(k) \in C^2$ and initial data $u(x, 0) = A_0(x)e^{ik_0x}$, where the function A_0 has the Fourier transform $\widehat{A}_0$ with compact support

$$\widehat{A}_0(k) = 0 \quad (|k| > r > 0), \quad M = \int_{-r}^{r} |\widehat{A}_0(k)|\,dk < \infty$$

(the initial perturbation with narrowband spectrum). Prove that for small $t > 0$ the solution u has the form of a modulated harmonic wave:

$$u(x, t) = A_0(x - \omega'(k_0)t)e^{i(k_0x - \omega(k_0)t)} + \widetilde{u}(x, t),$$

where the function $\widetilde{u}$ satisfies the uniform estimate with respect to x

$$|\widetilde{u}(x, t)| \leq \frac{1}{2}Mr^2t \max_{|k-k_0|\leq r} |\omega''(k)|.$$

22. Using the stationary phase method, find the asymptotics of the integral

$$v(t) = \int_{-\infty}^{+\infty} e^{i(tk - \frac{1}{3}k^3)}e^{-k^2/t}\,dk$$

as $t \to +\infty$.

Answer:

$$v(t) = \frac{2\sqrt{\pi}}{e\sqrt[4]{t}} \cos\left(\frac{2t\sqrt{t}}{3} - \frac{\pi}{4}\right) + O\left(\frac{1}{t}\right).$$

23. Using the representation of the Airy function as a contour integral in the plane of complex variable $\zeta = k + il$,

$$Ai(x) = \frac{1}{2\pi} \int_C e^{i(x\zeta + \frac{1}{3}\zeta^3)}\, d\zeta,$$

along the contour $C = \{k + il : l = (1/\sqrt{3})|k|\}$, show that the function $Ai(z)$ is an entire analytic function of the complex variable $z = x + iy$ and satisfies the differential equation

$$Ai''(z) = zAi(z).$$

24. We look for a self-similar solution $u(x, t) = (1/\sqrt[3]{3t})v(x/\sqrt[3]{3t})$ to the linear Korteweg-de Vries equation

$$u_t + u_{xxx} = 0.$$

What differential equation does the function $v(\xi)$ satisfy? Show that the Airy function $v = Ai(\xi)$ yields one of such solutions.
Answer: $v'' = \xi v + C$ ($C = \text{const}$).

25. Consider the Cauchy problem for the linearized Korteweg-de Vries equation

$$u_t + c_0 u_x + u_{xxx} = 0,$$
$$u(x, 0) = u_0(x),$$

where the initial function u_0 satisfies the condition

$$M \stackrel{\text{def}}{=} \int_{-\infty}^{+\infty} u_0(x)dx \neq 0.$$

Using the stationary phase method, find an asymptotic expansion of the solution in a neighborhood of the point $x = c_0 t$ ($c_0 > 0$) as $t \to +\infty$.
Answer:

$$u(x, t) = \frac{M}{\sqrt[3]{3t}} Ai\left(\frac{x - c_0 t}{\sqrt[3]{3t}}\right) + O\left(t^{-\frac{2}{3}}\right).$$

26. Find a general form of the scaling $\tilde{x} = ax, \tilde{t} = bt, \tilde{u} = cu$ reducing the Korteweg-de Vries equation

$$u_t + \alpha u u_x + \beta u_{xxx} = 0$$

where $\alpha, \beta = \text{const}, \alpha \neq 0, \beta \neq 0$, to the form (2.13) in variables $\tilde{x}, \tilde{t}, \tilde{u}$. What scaling of the variables x, t, u leaves the Korteweg-de Vries equation invariant?
Answer: $\tilde{x} = ax, \tilde{t} = \beta a^3 t, \tilde{u} = (\alpha/\beta)a^{-2}u$ ($a \in \mathbb{R} : a \neq 0$). The equation is invariant under the transformation $\tilde{x} = ax, \tilde{t} = a^3 t, \tilde{u} = a^{-2}u$.

27. For the Korteweg-de Vries equation written as

$$u_t + 6uu_x + u_{xxx} = 0 \tag{2.18}$$

construct a solitary wave solution $u = u(x - ct)$ ($u, u', u'' \to 0$ as $|x| \to \infty$) such that the wave crest at time $t = 0$ is located at a given point $x = x_0$.
Answer:

$$u(x, t) = \frac{c}{2 \cosh^2 \frac{\sqrt{c}}{2}(x - x_0 - ct)} \qquad (c > 0).$$

28. Consider the n-soliton solution to the Korteweg-de Vries equation (2.18)

$$u(x, t) = 2\frac{\partial^2}{\partial x^2} \ln \det \{I + A(x, t)\},$$

where I is the identity matrix of order n and A is a symmetric matrix with entries

$$a_{ij}(x, t) = \frac{\gamma_i \gamma_j}{k_i + k_j} e^{-(k_i + k_j)x} e^{4(k_i^3 + k_j^3)t} \qquad (i, j = 1, \ldots, n),$$

$k_1, \ldots, k_n$ and $\gamma_1, \ldots, \gamma_n$ ($k_i + k_j \neq 0$) are real constants.

a. What choice of constants γ_1 and k_1 in the solitary solution ($n = 1$) corresponds to a travelling wave type solitary solution (cf. Problem 27)?
b. Write explicitly a two-soliton solution ($n = 2$) in terms of hyperbolic functions with parameters $k_1 = \gamma_1 = \gamma_2 = 1$ and $k_2 = 2$.

Answer:

a. $\gamma_1^2 = \sqrt{c}e^{\sqrt{c}x_0}, k_1 = \sqrt{c}/2.$
b. $u(x, t) = 12\dfrac{3 + 4\cosh(2x - 8t) + \cosh(4x - 64t)}{\{3\cosh(x - 28t) + \cosh(3x - 36t)\}^2}.$

29. Consider the eigenvalue problem for the second order ordinary differential operator (the Sturm–Liouville problem)

$$\frac{d^2\varphi}{dx^2} + (u(x,t) + \lambda)\varphi = 0 \quad (-\infty < x < \infty)$$

with a given coefficient u depending on the parameter t (time). Show that if the function $u(x,t)$ is a solution to the Korteweg-de Vries equation (2.18), then the eigenfunction φ and eigenvalue λ satisfy the relation

$$\varphi^2 \frac{d\lambda}{dt} + (\varphi Q_x - \varphi_x Q)_x = 0,$$

where $Q = \varphi_t - \varphi_{xxx} + 3(u - \lambda)\varphi_x$. Show that $d\lambda/dt = 0$ if the functions φ and u, together with their derivatives, converge to zero as $|x| \to +\infty$.

30. Show that for solutions $u(x,t)$ of the Korteweg-de Vries equation (2.13) decaying, together with their derivatives, as $|x| \to \infty$, the following conservation laws hold:

$$\frac{d}{dt} \int_{-\infty}^{+\infty} u^2(x,t)dx = 0,$$

$$\frac{d}{dt} \int_{-\infty}^{+\infty} \left(u_x^2(x,t) - \frac{1}{3}u^3(x,t) \right) dx = 0.$$

31. We say that a differential equation with two independent variables x and t that is of the first order with respect to t admits the *Hamiltonian formulation* if it can be represented as

$$u_t = D_x \delta_u H(u, u_x, \ldots, u_{x\ldots x}^{(n)}).$$

Here, D_x is the operator of total differentiation with respect to x and δ_u is the Euler operator (the operator of variational differentiation),

$$D_x = \partial_x + u_x \partial_u + u_{xx} \partial_{u_x} + u_{xxx} \partial_{u_{xx}} + \ldots,$$

$$\delta_u = \partial_u - D_x \partial_{u_x} + D_x^2 \partial_{u_{xx}} - \ldots,$$

where $x, u, u_x, u_{xx}, \ldots$ are regarded as independent variables. Find the Hamilton function $H(u, u_x)$ for the Korteweg-de Vries equation (2.13).
Answer:

$$H(u, u_x) = \frac{1}{2}u_x^2 - \frac{1}{6}u^3.$$

32. Show that for any smooth function $v(x, t) \in C^4$ the pair of functions v and $u = -v_x - v^2$ is connected by the identity (the Gardner–Miura transformation)

$$u_t + 6uu_x + u_{xxx} = -(2v + \partial_x)(v_t - 6v^2 v_x + v_{xxx}).$$

33. For the modified Korteweg-de Vries equation

$$v_t - (6v + 6v^2)v_x + v_{xxx} = 0 \qquad (2.19)$$

find a bounded travelling wave type solution $v = v(x - t)$ such that $v, v', v'' \to 0$ as $x \to -\infty$.
Answer:

$$v = -\frac{1}{1 + e^{-(x-t)}} \quad \text{(a front type wave)}$$

34. Construct a travelling solitary wave type solution $v = v(x - ct)$ to Eq. (2.19) with a velocity $0 < c < 1$ such that $v, v', v'' \to 0$ as $x \to \pm\infty$.
Answer:

$$v = -\frac{c}{1 + \sqrt{1 - c}\cosh\sqrt{c}(x - ct)}.$$

35. Derive the dispersion relation for wave packets

$$\psi(\mathbf{x}, t) = a\exp(i\mathbf{k} \cdot \mathbf{x} - \omega t)$$

satisfying the Schrödinger equation with constant potential $V = \text{const}$:

$$i\hbar\psi_t + \frac{\hbar^2}{2m}\Delta\psi - V\psi = 0,$$

where $\Delta = \partial_x^2 + \partial_y^2 + \partial_z^2$ is the Laplace operator, $\hbar$ is the Planck constant (the de Broglie wave in quantum mechanics describing the behavior of a particle of mass m in a field with potential V). Show that the particle velocity $\mathbf{U} = \mathbf{p}/m$, defined as the ratio of its momentum $\mathbf{p} = \hbar\mathbf{k}$ ($\mathbf{k}$ is the wave vector) to the mass m, coincides with the group velocity $\mathbf{c}_g(\mathbf{k})$.

36. Consider the nonlinear Schrödinger equation

$$i\hbar\psi_t + \frac{\hbar^2}{2m}\Delta\psi - V(|\psi|^2)\psi = 0$$

with nonlinear potential $V = V(|\psi|^2)$. Show that, under the change of the sought complex-valued functions

$$\psi(x, t) = \sqrt{\rho(x, t)}e^{i\theta(x,t)/\hbar}, \quad \mathbf{v} = \nabla\theta,$$

where the functions $\rho \geqslant 0$ and θ are real-valued (the Madelung transformation), this equation goes to the following system of equations for ρ and $\mathbf{v}$:

$$\rho_t + \operatorname{div}(\rho \mathbf{v}) = 0,$$

$$m(\mathbf{v}_t + (\mathbf{v} \cdot \nabla)\mathbf{v}) + \nabla \left(V(\rho) - \frac{\hbar^2}{2m} \frac{\Delta(\sqrt{\rho})}{\sqrt{\rho}} \right) = 0. \tag{2.20}$$

37. Show that the system (2.20) in the one-dimensional case takes the form

$$\rho_t + (\rho v)_x = 0,$$

$$m(v_t + v v_x) + \left\{ V(\rho) - \frac{\hbar^2}{4m} \left(\frac{\rho_{xx}}{\rho} - \frac{\rho_x^2}{2\rho^2} \right) \right\}_x = 0.$$

Linearize this system at the constant solution $\rho = \rho_0$, $v = 0$ and derive the dispersion relation. Show that the wave modes are real for all wavenumbers if and only if $V'(\rho) \geqslant 0$.

38. Consider the nonlinear Schrödinger equation

$$i\psi_t + \gamma \psi_{xx} + \beta |\psi|^2 \psi = 0 \quad (\gamma, \beta \in \mathbb{R} : \gamma \neq 0, \ \beta \neq 0). \tag{2.21}$$

Verify that this equation is invariant under the transformation $t \to -t$, $\psi \to \overline{\psi}$ (the bar means complex conjugation). Find all homothetic transformations $\widetilde{x} = ax$, $\widetilde{t} = bt$, $\widetilde{\psi} = c\psi$, where a, b, and c are real parameters, that leave Eq. (2.21) invariant. Show that, under a suitable homothetic transformation of the variables x, t, ψ, Eq. (2.21) takes one of the following two canonical forms:

$$i\psi_t + \psi_{xx} \pm |\psi|^2 \psi = 0.$$

To which form does the equation reduce? What is the most general form of the reducing transformation?

Answer: The sign "plus" if $\gamma\beta > 0$ and "minus" if $\gamma\beta < 0$; $\widetilde{x} = ax$, $\widetilde{t} = \gamma a^2 t$, $\widetilde{\psi} = \pm\sqrt{|\beta/\gamma|}a^{-1}\psi$.

39. Show that for solutions to the nonlinear Schrödinger equation (2.11) decaying as $|x| \to \infty$ the following conservation laws hold:

$$\frac{d}{dt} \int_{-\infty}^{+\infty} |\psi(x, t)|^2 dx = 0,$$

$$\frac{d}{dt} \int_{-\infty}^{+\infty} \left(|\psi_x(x, t)|^2 \mp \frac{1}{2}|\psi(x, t)|^4 \right) dx = 0.$$

40. For the nonlinear Schrödinger equation (2.11) with the sign "minus"

$$i\psi_t + \psi_{xx} - |\psi|^2\psi = 0$$

construct a solution of the form $\psi(x,t) = A(x-ct)\exp i(kx - \omega t)$ with the real amplitude A such that $|A| \to a = \text{const} > 0$, $A' \to 0$ as $|x| \to \infty$.
Answer:

$$\psi(x,t) = \pm a \tanh\left\{\frac{a(x-ct)}{\sqrt{2}}\right\} e^{i(kx-\omega t)},$$

where $\omega = a^2 + k^2$, $c = 2k$.

41. Find the general form of the self-similar solution $u(x,t) = v(x/\sqrt{t})/\sqrt{t}$ for the one-dimensional Schrödinger equation

$$iu_t + u_{xx} = 0.$$

Answer:

$$u(x,t) = \frac{1}{\sqrt{t}} e^{ix^2/4t}\left(C_1 + C_2 \int\limits_0^{x/\sqrt{t}} e^{-i\xi^2/4}\, d\xi\right) \quad (C_1, C_2 \in \mathbb{C}).$$

42. Using the Fourier transform, construct a solution to the Cauchy problem

$$iu_t + u_{xx} = 0,$$

$$u(x,0) = \frac{\sin x}{x}.$$

Answer:

$$u(x,t) = \frac{1}{2\sqrt{t}} e^{\frac{ix^2}{4t}}\left\{f\left(\frac{2t-x}{2\sqrt{t}}\right) + f\left(\frac{2t+x}{2\sqrt{t}}\right)\right\}, \quad f(z) = \int\limits_0^z e^{-is^2}\, ds.$$

43. Let a complex-valued function $u(x,t) \in C^4$ be a solution to the Cauchy problem for the linear Schrödinger equation

$$iu_t + \gamma u_{xx} = 0,$$
$$u(x,0) = u_0(x),$$

where γ is a constant and the initial function u_0 takes only real values, $\text{Im}\, u_0(x) \equiv 0$. Prove that the function $v(x,t) = \text{Re}\, u(x,t)$ is a solution the

Cauchy problem for the equation of flexural waves in an elastic rod

$$v_{tt} + \gamma^2 v_{xxxx} = 0,$$
$$v(x, 0) = u_0(x), \quad v_t(x, 0) = 0.$$

44. Using the Fourier transform, construct explicitly a solution to the Euler–Bernoulli beam equation

$$u_{tt} + \gamma^2 u_{xxxx} = 0 \quad (-\infty < x < +\infty),$$

with the initial data

$$u(x, 0) = ae^{-x^2/4}, \quad u_t(x, 0) = 0 \quad (a = \text{const})$$

Hint: To take the inverse Fourier transform, use the formula

$$\int_{-\infty}^{+\infty} e^{-\zeta k^2} e^{ikx} dk = \sqrt{\frac{\pi}{\zeta}} e^{-\frac{x^2}{4\zeta}} \quad (\zeta \in \mathbb{C} : \text{Re}\,\zeta > 0).$$

Choose the branch for which $\sqrt{1} = 1$.
Answer:

$$u(x, t) = \frac{a}{\sqrt[4]{1 + \gamma^2 t^2}} e^{-\frac{x^2}{4(1+\gamma^2 t^2)}} \cos\left\{\frac{\gamma t x^2}{4(1 + \gamma^2 t^2)} - \frac{1}{2} \arctan \gamma t\right\}.$$

45. Derive the dispersion relation, find the normal phase and group velocities for the equation of bending oscillations of a plate

$$u_{tt} + \gamma^2 \Delta^2 u = 0,$$

where $\Delta = \partial_{x_1}^2 + \partial_{x_2}^2$ is the two-dimensional Laplace operator, $\gamma^2 = Eh^2/(12\rho_0(1 - \nu^2))$, h is the plate thickness, E is the Young modulus, and ν is the Poisson coefficient.
Answer: $\omega(\mathbf{k}) = \pm\gamma|\mathbf{k}|^2$, $c_p(\mathbf{k}) = \pm\gamma|\mathbf{k}|$, $c_g(\mathbf{k}) = \pm2\gamma\mathbf{k}$.

46. For the displacement vector $\mathbf{w} = (w_1, w_2, w_3)^T$ solving the three-dimensional Lamé system of equations

$$\rho_0 \mathbf{w}_{tt} = (\lambda + \mu)\nabla \operatorname{div} \mathbf{w} + \mu\Delta\mathbf{w}$$

consider the Helmholtz representation

$$\mathbf{w} = \nabla\varphi + \operatorname{curl}\mathbf{v} \quad (\operatorname{div}\mathbf{v} = 0),$$

where φ and $\mathbf{v}$ are defined for all $\mathbf{x} = (x_1, x_2, x_3)^T \in \mathbb{R}^3$ and have bounded second order derivatives with respect to $\mathbf{x}$ and t. Show that the scalar potential φ and vector potential $\mathbf{v}$ satisfy the wave equations

$$\varphi_{tt} = c_1^2 \Delta \varphi,$$

$$\mathbf{v}_{tt} = c_2^2 \Delta \mathbf{v}.$$

What are the propagation velocities c_1 and c_2 of the corresponding waves?
Answer:

$$c_1 = \sqrt{\frac{\lambda + 2\mu}{\rho_0}} \text{ is the velocity of longitudinal waves}$$

$$c_2 = \sqrt{\frac{\mu}{\rho_0}} \text{ is the velocity of transverse waves.}$$

47. The equality (2.6) connecting the amplitude vector $\mathbf{a}$, wave vector $\mathbf{k}$, and frequency ω of the wave packet for the three-dimensional Lamé system of equations can be written as the linear system of equations

$$A(\omega, \mathbf{k})\mathbf{a} = 0$$

which is homogeneous with respect to $\mathbf{a}$, where

$$A(\omega, \mathbf{k}) = (\rho_0 \omega^2 - \mu |\mathbf{k}|^2)I - (\lambda + \mu)\mathbf{k} \otimes \mathbf{k},$$

the symbol $\otimes$ means the tensor product of vectors. Find the dispersive function $D(\omega, \mathbf{k}) = \det A(\omega, \mathbf{k})$ for the Lamé equations.
Answer:

$$D(\omega, \mathbf{k}) = (\rho_0 \omega^2 - \mu |\mathbf{k}|^2)^2 [\rho_0 \omega^2 - (\lambda + 2\mu)|\mathbf{k}|^2].$$

48. Using the two-dimensional Lamé system of equations with displacement vector $\mathbf{w} = (w_1, w_2)^T$, find the phase velocity c_p of the Rayleigh surface waves described by the wave packets

$$w_1 = A_1(x_2) \cos(kx_1 - \omega t),$$

$$w_2 = A_2(x_2) \sin(kx_1 - \omega t)$$

with the stress-free condition

$$\partial_{x_2} w_1 + \partial_{x_1} w_2 = 0,$$

$$\lambda \partial_{x_1} w_1 + (\lambda + 2\mu)\partial_{x_2} w_2 = 0$$

on the boundary of the half-spaces $x_2 \leqslant 0$ and the decay condition

$$\mathbf{w} \to 0 \quad \text{as } x_2 \to -\infty.$$

Show that the phase velocity c_p of the Rayleigh waves satisfies the inequalities $c_1 > c_2 > c_p$, where c_1 is the velocity of longitudinal waves and c_2 is the velocity of transverse waves (cf. Problem 46).
Answer: $c_p = \sqrt{s_0}c_2$, where s_0 is the least real root of the equation

$$s^3 - 8s^2 + 16(3/2 - (c_2/c_1)^2)s - 16(1 - (c_2/c_1)^2) = 0.$$

49. Small oscillations of identical masses linked by springs of rigidity β^2 are described by the system of ordinary differential equations

$$m\ddot{w}_n = \beta^2(w_{n+1} - 2w_n + w_{n-1}), \quad n \in \mathbb{Z}.$$

Find the phase and group velocities of the signal propagation along the chain in the form of the wave packet $w_n(t) = a \exp\{i(kn - \omega t)\}$ (k is the dimensionless wavenumber).
Answer:

$$c_p(k) = \pm \frac{\beta}{\sqrt{m}} \frac{\sin(k/2)}{(k/2)}, \quad c_g(k) = \pm \frac{\beta}{\sqrt{m}} \cos(k/2).$$

50. Under the conditions of the previous problem, show that for those motions of masses that satisfy the decay condition $w_n, \dot{w}_n \to 0$ ($n \to \pm\infty$) the total energy E of the system is preserved with time:

$$E(t) \stackrel{\text{def}}{=} \frac{1}{2} \sum_{n=-\infty}^{+\infty} \{m\dot{w}_n^2 + \beta^2(w_{n+1} - w_n)^2\} = \text{const} .$$

51. Write an equation of small oscillations and deduce the dispersion relation for a chain with alternating masses $m_1 \neq m_2$ linked by identical springs of rigidity β^2. What is the dispersion relation in the limit of identical masses $m_1 = m_2 = m$?
Answer:

$$\omega^2(k) = \beta^2 \left\{ \frac{1}{m_1} + \frac{1}{m_2} \pm \sqrt{\left(\frac{1}{m_1} - \frac{1}{m_2}\right)^2 + \frac{4\cos^2 k}{m_1 m_2}} \right\}.$$

52. Consider a nonlinear chain of elastic beads of the same mass and radius R that interact according to the Hertz law:

$$F_n = \frac{\sqrt{2R^2 E}}{3(1 - v^2)} \left\{ 2 - \frac{x_{n+1} - x_n}{R} \right\}^{3/2},$$

where E is the Young modulus, v is the Poisson coefficient, $x_{n+1} > x_n$ are the centers of neighboring beads, and F_n is the force of their elastic interaction. The equations describing displacements of the bead centers w_n in dimensionless variables is written as

$$\ddot{w}_n = (\delta - w_n + w_{n-1})^{3/2} - (\delta - w_{n+1} + w_n)^{3/2}, \quad n \in \mathbb{Z},$$

where the parameter $\delta \geqslant 0$ characterizes the initial displacement of the bead centers. Linearize the equation of chain in the case $\delta \neq 0$ and deduce the dispersion relation.

53. Consider the equation

$$u_{tt} = (u^{3/2} + u^{1/4}(u^{5/4})_{xx})_{xx}$$

governing the propagation of one-dimensional nonlinear waves in weakly compressible granular materials (here, $u = -w_x > 0$ is the strain and w is the displacement). This equation corresponds to the long-wave approximation for a chain of elastic beads interacting in accordance with the Hertz law (cf. the previous problem) without initial displacement ($\delta = 0$). Verify that this equation has an exact travelling wave type solution

$$u_c(x, t) = \frac{25}{16} c^4 \cos^4 \left(\frac{x - ct}{5} \right).$$

Show that the function $\eta(x, t)$ defined by the conditions $\eta = u_c$ for $|x - ct| < (5/2)\pi$ and $\eta = 0$ for $|x - ct| \geqslant (5/2)\pi$ is also a solution (a compacton, i.e., a solitary wave with compact support).

54. Derive the dispersion relation for the linearized Whitham integro-differential equation

$$u_t + \frac{\pi}{4} \int_{-\infty}^{+\infty} e^{-\frac{\pi}{2}|x-y|} u_y(y, t) dy = 0.$$

Answer:

$$\omega(k) = \frac{\pi^2 k}{\pi^2 + 4k^2}.$$

55. The Hilbert integral transform H acts on a function u by the formula

$$Hu(x) = v.p. \frac{1}{\pi} \int_{-\infty}^{+\infty} \frac{u(y)dy}{x-y} = \lim_{\substack{A \to +\infty \\ \varepsilon \to 0+}} \frac{1}{\pi} \left(\int_{x-A}^{x-\varepsilon} + \int_{x+\varepsilon}^{x+A} \right) \frac{u(y)dy}{x-y}.$$

How are the Fourier transform $\widehat{Hu}(k)$ of $Hu(x)$ and the Fourier transform of $u(x)$ connected? Find the Hilbert transform $Hv(x)$ of the function $v(x) = b/(x^2 + b^2)$ ($b > 0$).
Answer: $\widehat{Hu}(k) = -i\,\mathrm{sgn}\,k\,\widehat{u}(k)$, $Hv(x) = x/(x^2 + b^2)$.

56. Looking for a solution in the form of an elementary wave packet $u(x,t) = ae^{i(kx-\omega t)}$, deduce the dispersion relation for the linearized Benjamin–Ono equation

$$u_t + c_0 u_x + H u_{xx} = 0,$$

where H is the Hilbert transform.
Answer: $\omega(k) = (c_0 + |k|)k$.

57. Construct a solution to the Benjamin–Ono equation

$$u_t + u u_x + H u_{xx} = 0$$

in the form of a travelling solitary wave $u = u(x - ct)$, where the fractional-rational function $u(\xi)$ is a linear combination of the functions $v_1(\xi) = b/(\xi^2 + b^2)$ and $v_2(\xi) = \xi/(\xi^2 + b^2)$ (use the result of Problem 55 and the fact that the Hilbert transform commutes with the operator of differentiation).
Answer:

$$u(x,t) = \frac{4c}{c^2(x - ct)^2 + 1}.$$

Chapter 3
Water Waves

3.1 Equations of Motion

We consider the motion of an ideal incompressible inhomogeneous fluid in the gravity field $\mathbf{g} = (0, 0, -g)^T$. The unknowns are the velocity vector $\mathbf{u} = (u, v, w)^T$, the density ρ, and the pressure p depending on $\mathbf{x} = (x, y, z)^T \in \mathbb{R}^3$ and the time t. The integral laws of conservation of volume

$$\iint_S \mathbf{u} \cdot \mathbf{n}\, dS = 0,$$

mass

$$\frac{d}{dt} \iiint_\Omega \rho\, d\Omega + \iint_S \rho \mathbf{u} \cdot \mathbf{n}\, dS = 0,$$

momentum

$$\frac{d}{dt} \iiint_\Omega \rho \mathbf{u}\, d\Omega + \iint_S (\rho \mathbf{u}(\mathbf{u} \cdot \mathbf{n}) + p\mathbf{n})\, dS = \iiint_\Omega \rho \mathbf{g}\, d\Omega,$$

and energy

$$\frac{d}{dt} \iiint_\Omega \left(\frac{1}{2}\rho|\mathbf{u}|^2 + \rho g z \right) d\Omega + \iint_S \left(\frac{1}{2}\rho|\mathbf{u}|^2 + p + \rho g z \right) (\mathbf{u} \cdot \mathbf{n})\, dS = 0$$

hold in any fixed domain Ω with piecewise smooth boundary S ($\mathbf{n}$ is the unit outward normal to S). In the domain, where the motion is described by smooth functions $\mathbf{u}$,

© Springer International Publishing AG 2017
S.L. Gavrilyuk et al., *Waves in Continuous Media*, Lecture Notes in Geosystems
Mathematics and Computing, DOI 10.1007/978-3-319-49277-3_3

ρ, and p, the above-mentioned set of conservation laws is equivalent to the system of differential equations

$$\text{div } \mathbf{u} = 0,$$

$$\rho_t + \mathbf{u} \cdot \nabla \rho = 0, \tag{3.1}$$

$$\mathbf{u}_t + (\mathbf{u} \cdot \nabla)\mathbf{u} + \frac{1}{\rho}\nabla p = \mathbf{g}.$$

Moreover, the vorticity $\boldsymbol{\omega} = \text{curl } \mathbf{u}$ satisfies the Helmholtz equation

$$\boldsymbol{\omega}_t + (\mathbf{u} \cdot \nabla)\boldsymbol{\omega} = (\boldsymbol{\omega} \cdot \nabla)\mathbf{u} - \frac{1}{\rho^2}\nabla p \times \nabla \rho.$$

Hence vorticity in a nonviscous inhomogeneous fluid changes under the influence of two factors: transfer of the initial vorticity distribution and formation of new vorticity caused by the noncoincidence of level surfaces for the pressure and density in the flow (the isobaric surfaces $p(\mathbf{x}, t) = \text{const}$ and the isochoric surfaces $\rho(\mathbf{x}, t) = \text{const}$).

In the case of a constant density $\rho = \rho_0$, the solutions to the system (3.1) describe homogeneous fluid flows. In this case, the system is reduced to

$$\text{div } \mathbf{u} = 0,$$

$$\mathbf{u}_t + (\mathbf{u} \cdot \nabla)\mathbf{u} + \frac{1}{\rho_0}\nabla p = \mathbf{g}. \tag{3.2}$$

Respectively, the Helmholtz equation takes the form

$$d_t\boldsymbol{\omega} = \frac{\partial \mathbf{u}}{\partial \mathbf{x}}\langle \boldsymbol{\omega} \rangle, \quad d_t = \partial_t + (\mathbf{u} \cdot \nabla). \tag{3.3}$$

Due to the special structure, the system (3.3) is integrated by passing from the Eulerian coordinates $(\mathbf{x}, t)$ to the Lagrangian coordinates $(\boldsymbol{\xi}, t)$. The dependence $\mathbf{x} = \mathbf{x}(\boldsymbol{\xi}, t)$ between the Eulerian and Lagrangian coordinates is determined by the solution to the Cauchy problem for differential equations for trajectories of fluid particles

$$\frac{d\mathbf{x}}{dt} = \mathbf{u}(\mathbf{x}, t),$$

$$\mathbf{x}_{|t=0} = \boldsymbol{\xi}. \tag{3.4}$$

In the Lagrangian coordinates, for the sought function $\widetilde{\boldsymbol{\omega}}(\boldsymbol{\xi}, t) = \boldsymbol{\omega}(\mathbf{x}(\boldsymbol{\xi}, t), t)$ we have $\partial_t\widetilde{\boldsymbol{\omega}} = d_t\boldsymbol{\omega}$. Therefore, for each trajectory with a given initial location of a particle $\boldsymbol{\xi}$ the system (3.3) is a system of ordinary differential equations for the vector-valued function $\widetilde{\omega}(\xi, t)$. On the other hand, differentiating (3.4) with respect to the parameter ξ, we obtain the equation in variations

$$\partial_t M = \frac{\partial \mathbf{u}}{\partial \mathbf{x}} \circ M, \quad M_{|t=0} = I$$

with the Jacobi matrix

$$M = \mathbf{x}'_\xi(\xi, t) = \frac{\partial(x, y, z)}{\partial(\xi, \eta, \zeta)}$$

and the identity matrix I. By the Ostrogradskii–Liouville formula, for the determinant $|M| = \det M$ we have

$$|M|_t = |M| tr \mathbf{u}'_\mathbf{x} = |M| \operatorname{div} \mathbf{u} = 0.$$

Consequently, the matrix M is nonsingular: $|M(\xi, t)| = 1$. Hence M is the fundamental matrix of solutions to the linear system (3.3) of ordinary differential equations. Thus, the variation of vorticity along trajectories of fluid particles is described by the Cauchy formula

$$\boldsymbol{\omega} = M \boldsymbol{\omega}_0,$$

where $\boldsymbol{\omega}_0$ is the initial vorticity field. As a consequence, we arrive at the Lagrange theorem asserting that $\boldsymbol{\omega} \equiv 0$ in a volume of a homogeneous fluid $\Omega(t)$ for all $t > 0$ if the vorticity vanishes in $\Omega(0)$. For a potential flow the velocity vector field $\mathbf{u}$ has a potential φ such that $\mathbf{u} = \nabla\varphi$. The function φ is harmonic with respect to the spatial variables x, y, z ($\Delta\varphi = 0$). In this case, the momentum equation in the system (3.2) is reduced to the Cauchy–Lagrange integral

$$\varphi_t + \frac{1}{2}|\nabla\varphi|^2 + \frac{1}{\rho_0}p + gz = b(t),$$

where b is an arbitrary function.

To determine uniquely the unsteady motion in the whole domain occupied by a fluid, we set $\mathbf{u} = \mathbf{u}_0$ at $t = 0$ and impose boundary conditions at the boundary of the domain for $t > 0$, in particular, the impermeability condition $\mathbf{u} \cdot \mathbf{n} = 0$ at the fixed boundary part with normal vector $\mathbf{n}$ and the kinematic condition

$$(f_t + \mathbf{u} \cdot \nabla f)_{|f=0} = 0$$

together with the dynamic condition

$$p = \widetilde{p},$$

where $\widetilde{p}$ is a given function, at the free boundary $f(x, y, z, t) = 0$. For example, for a fluid in contact with the atmosphere the pressure at the free surface is assumed to be constant if the air motion is not taken into account: $p = p_0$. In this case, without loss of generality we can assume that $p_0 = 0$. The kinematic condition at the interface of two immiscible fluids has the form

$$(f_t + \mathbf{u}_j \cdot \nabla f)|_{f=0} = 0 \quad (j = 1, 2),$$

where $\mathbf{u}_j$ are the limiting velocity values from both sides of the contact surface. If the surface tension is not taken into account, then the dynamic condition implies the pressure continuity

$$[p] = p_2 - p_1 = 0.$$

Stationary flows are described by solutions to the system (3.1) such that $\rho_t = 0$ and $\mathbf{u}_t = 0$. In this case, for any stream line

$$L : \frac{dx}{u} = \frac{dy}{v} = \frac{dz}{w}$$

the Bernoulli integral holds:

$$\frac{1}{2}|\mathbf{u}|^2 + \frac{1}{\rho}p + gz = b(L)$$

(the Bernoulli constant $b(L)$ depends on the stream line).

Example 3.1 Consider the plane flow of a two-layer fluid simulating the process of penetrating the bottom layer of a heavy fluid into a light fluid (cf. Fig. 3.1). We assume that the fluid densities are constant and are equal to ρ_1 and ρ_2 respectively ($\rho_2 < \rho_1$). The flow is stationary in the reference frame moving with the point of contact of the interface and the bottom. It is additionally assumed that, in this moving reference frame, the heavy fluid is at rest (this flow scheme is due to [17].

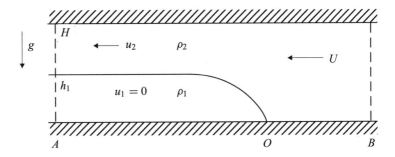

Fig. 3.1 A gravity current head of a heavy fluid penetrating a light fluid

In Fig. 3.1, A and B denote the limiting flow states with constant velocities as $x \to -\infty$ and $x \to +\infty$ respectively. The law of conservation of mass for the upper layer is written as

$$UH = u_2(H - h_1).$$

The law of conservation of total horizontal momentum of the fluid has the form

$$\int_0^H p_A \, dy + \int_{h_1}^H p_2 u_2^2 \, dy = \int_0^H \left(p_B + p_2 U^2\right) dy. \tag{3.5}$$

By the vertical momentum equation, the pressure at the state B is distributed hydrostatically

$$p_B(y) = p_B|_{y=0} - g\rho_2 y.$$

Using the Bernoulli integral on the bottom part OB, we get

$$p_B|_{y=0} = p_O - \frac{1}{2}\rho_2 U^2.$$

Similarly, applying the Bernoulli integral on the bottom part AO, we see that the pressure is constant: $p = p_O$. From the vertical momentum equation for the state A we obtain the hydrostatic distribution of pressure

$$p_A(y) = \begin{cases} p_O - g\rho_1 y, & 0 < y < h_1 \\ p_O - g\rho_1 h_1 - g\rho_2(y - h_1), & h_1 < y < H. \end{cases}$$

Substituting the obtained expressions for pressure distributions into (3.5) and removing u_2 by using the law of conservation of mass, we find

$$\frac{U^2}{gH} = \frac{\rho_1 - \rho_2}{\rho_2} \frac{(2H - h_1)(H - h_1)h_1}{H^2(H + h_1)}.$$

On the other hand, applying the Bernoulli integral to the upper fluid along the interface, we have

$$p_A|_{y=h_1} + \frac{1}{2}\rho_2 u_2^2 + g\rho_2 h_1 = p_O.$$

Using again the law of conservation of mass and the known expression for $p_A(y)$, we find

$$\frac{U^2}{gH} = \frac{2(\rho_1 - \rho_2)}{\rho_2} \frac{(H - h_1)^2 h_1}{H^3}.$$

Comparing these two different expressions for U^2/gH, we conclude that $h_1 = H/2$. Thus, the laws of conservation of mass, momentum, and energy simultaneously hold for the flow configuration under consideration only if the depth of the lower layer is equal to half the total depth of the channel.

3.2   Linear Theory of Surface Waves

We consider the potential flow of a homogeneous fluid in the layer $\Omega(t) = \{(x, y) \in \mathbb{R}^2 : 0 < z < h(x, y, t)\}$ bounded by the free surface $z = h(x, y, t)$ and the flat bottom $z = 0$ (cf. Fig. 3.2).

The equations for the velocity potential φ and function h have the form

$$\Delta\varphi \equiv \varphi_{xx} + \varphi_{yy} + \varphi_{zz} = 0, \quad \mathbf{x} \in \Omega(t),$$

$$\varphi_z = 0, \quad z = 0,$$

$$\left.\begin{aligned}
h_t + \varphi_x h_x + \varphi_y h_y - \varphi_z &= 0, \\
\varphi_t + \frac{1}{2}(\varphi_x^2 + \varphi_y^2 + \varphi_z^2) + gh &= 0,
\end{aligned}\right\} \quad z = h(x, y, t).$$

The system of these equations with the initial condition

$$h = h^{(0)}(x, y), \quad \varphi = \varphi^{(0)}(x, y, z) \quad (\Delta\varphi^{(0)} = 0, \mathbf{x} \in \Omega(0))$$

is called the *Cauchy–Poisson problem*.

The equations under consideration have an exact solution $h = h_0 = \text{const}$, $\varphi = \varphi_0 = u_0 x - \frac{1}{2}u_0^2 t - gh_0 t$ describing the uniform motion of a fluid layer of constant depth h_0 with constant velocity u_0 directed along the Ox-axis. Small perturbations $\varphi = \varphi_0 + \Phi$, $h = h_0 + \zeta$ of a given state are approximately described by the linear

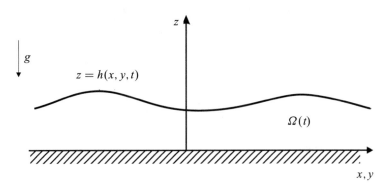

Fig. 3.2 A sketch for flow with free surface over a flat bottom

system of equations

$$\Phi_{xx} + \Phi_{yy} + \Phi_{zz} = 0 \quad (0 < z < h_0),$$

$$\Phi_z = 0, \quad z = 0,$$

$$\left.\begin{array}{c} \zeta_t + u_0 \zeta_x - \Phi_z = 0 \\ \Phi_t + u_0 \Phi_x + g\zeta = 0 \end{array}\right\} \quad z = h_0.$$

We consider solutions to the Laplace equation in the form of wave packets satisfying the impermeability condition at the bottom

$$\zeta = a e^{i(kx+ly-\omega t)}, \quad \Phi = b\,\mathrm{ch}\,mz e^{i(kx+ly-\omega t)}, \quad m = \sqrt{k^2 + l^2}.$$

From the boundary conditions at $z = h_0$ it follows that such solutions with nonzero amplitudes a and b exist if and only if the frequency ω and wave vector $\mathbf{k} = (k, l)^T$ satisfy the dispersion relation

$$(\omega - u_0 k)^2 = gm \tanh m h_0. \tag{3.6}$$

For plane waves we have $l = 0$ and $m = |k|$. In this case, the phase and group velocities are expressed by

$$c_p = \frac{\omega}{k} = u_0 \pm \sqrt{gh_0} \sqrt{\frac{\tanh kh_0}{kh_0}},$$

$$c_g = \frac{d\omega}{dk} = u_0 \pm \sqrt{gh_0} f(kh_0),$$

where

$$f(\xi) = \frac{d}{d\xi} \sqrt{\xi \tanh \xi} = \frac{1}{2}\left(\sqrt{\frac{\tanh \xi}{\xi}} + \frac{1}{\cosh^2 \xi} \sqrt{\frac{\xi}{\tanh \xi}} \right).$$

We emphasize the following particular cases.

(a) *Stationary waves.* In this case, the phase velocity of the wave vanishes: $c_p = 0$, and $F = \sqrt{\tanh kh_0/(kh_0)}$ for the *Froude number* $F = |u_0|/\sqrt{gh_0}$. The quantity $\sqrt{gh_0}$ is called the *critical velocity*. Hence we can assert that linear stationary waves can occur only for the subcritical flow with velocity $|u_0| < \sqrt{gh_0}$.

(b) *Deep water waves.* In the limit as $h_0 \to \infty$, the phase and group velocities of waves propagating in a fluid at rest ($u_0 = 0$) are expressed as $c_p = \sqrt{g/k}$ and $c_g = \frac{1}{2}\sqrt{g/k}$. Thus, for a deep water wave the group velocity is half the phase velocity.

(c) *Long waves.* In this limit case, we let $k \to 0$. Consequently, $c_p = u_0 \pm \sqrt{gh_0} = c_g$. The coincidence of the phase velocity and the group velocity indicates the hyperbolicity of the long waves.

Problem 3.1 Find trajectories of particles of a plane travelling wave with the velocity potential $\Phi = ga\omega^{-1}e^{ky}\sin(kx - \omega t)$, where ω and k are connected by the dispersion relation $\omega^2 = gk$ in the linear theory of deep water waves (it is assumed that the parameter $\alpha = ak$ is small).

Solution The potential $\Phi(x, y, t)$ is a harmonic function of x and y for $y < 0$, satisfies the decay condition $\nabla\Phi \to 0$ as $y \to -\infty$ and the boundary conditions $\zeta_t = \Phi_y$, $\Phi_t + g\zeta = 0$ on $y = 0$, where the function ζ defining the free surface profile $y = \zeta(x, t)$ has the form of a real wave packet $\zeta(x, t) = a\cos(kx - \omega t)$. The trajectories of particles with the velocity field $\mathbf{u} = \nabla\Phi$ are described by the differential equations

$$\frac{dx}{dt} = ga\omega^{-1}e^{ky}\cos(kx - \omega t),$$

$$\frac{dy}{dt} = ga\omega^{-1}e^{ky}\sin(kx - \omega t) \tag{3.7}$$

and the initial data $(x(0), y(0)) = (\xi, \eta)$. The smallness of the parameter $\alpha = ak = 2\pi a/L$ implies the smallness of the ratio of the wave amplitude a to the length L. Taking into account this property, we look for a solution $\mathbf{x} = (x, y)$ in the form

$$\mathbf{x}(t) = \mathbf{x}_0(t) + \alpha\mathbf{x}_1(t) + O(\alpha^2).$$

Substituting the last expression into Eqs. (3.7) and collecting terms at the same powers of α, we obtain the following equation for $\mathbf{x}_0(t)$ and $\mathbf{x}_1(t)$:

$$\frac{dx_0}{dt} = 0, \quad \frac{dy_0}{dt} = 0, \quad (x_0, y_0)|_{t=0} = (\xi, \eta),$$

$$\left.\begin{array}{l} \dfrac{dx_1}{dt} = \dfrac{\omega}{k}e^{ky_0}\cos(kx_0 - \omega t), \\[3mm] \dfrac{dy_1}{dt} = \dfrac{\omega}{k}e^{ky_0}\sin(kx_0 - \omega t), \end{array}\right\} \quad (x_1, y_1)|_{t=0} = (0, 0).$$

Hence $(x_0(t), y_0(t)) = (\xi, \eta)$ and, consequently,

$$x_1(t) = \frac{1}{k}e^{k\eta}(\sin k\xi - \sin(k\xi - \omega t)),$$

$$y_1(t) = \frac{1}{k}e^{k\eta}(\cos(k\xi - \omega t) - \cos k\xi).$$

Thus, the functions $\mathbf{x}_0(t)$ and $\mathbf{x}_1(t)$ are periodic with period $T = 2\pi/\omega$ equal to the time period of a progressive harmonic wave. For the period T the particle with coordinates $\mathbf{x}(t) = \mathbf{x}_0(t) + \alpha\mathbf{x}_1(t)$, taken up to a summand of order $O(\alpha^2)$, runs the circle $|\mathbf{x} - \mathbf{x}_c| = r$ with radius $r = ae^{k\eta}$ and center $\mathbf{x}_c = (\xi + ae^{k\eta}\sin k\xi, \eta - ae^{k\eta}\cos k\xi)$. Hence the oscillation amplitude of particles in the wave is maximal on the fluid surface and exponentially decreases with increasing the submersion depth. □

Answer:

$$x(t) = \xi + ae^{k\eta}(\sin k\xi - \sin(k\xi - \omega t)) + O(a^2 k^2),$$

$$y(t) = \eta - ae^{k\eta}(\cos k\xi - \cos(k\xi - \omega t)) + O(a^2 k^2).$$

3.3 Shallow Water Theory

In the theory of nonlinear long waves, it is assumed that the characteristic wave-length L is much larger than the fluid depth h_0. With these scales, we introduce the dimensionless variables x', t', φ', h':

$$(x, y) = L(x', y'), \quad (z, h) = h_0(z', h'), \quad t = \frac{L}{\sqrt{gh_0}}t', \quad \varphi = L\sqrt{gh_0}\varphi'.$$

Then the original equations with small parameter $\varepsilon = h_0/L$ take the form

$$\varepsilon^2(\varphi_{xx} + \varphi_{yy}) + \varphi_{zz} = 0 \quad (0 < z < h),$$

$$\varphi_z = 0 \quad (z = 0),$$

$$\left.\begin{array}{c} h_t + \varphi_x h_x + \varphi_y h_y = \varepsilon^{-2}\varphi_z \\ \\ \varphi_t + \dfrac{1}{2}(\varphi_x^2 + \varphi_y^2 + \varepsilon^{-2}\varphi_z^2) + h = 0 \end{array}\right\} \quad (z = h)$$

(we omit the primes in the notation of dimensionless quantities). The Lagrange method in the theory of long waves uses the representation of the potential φ in terms of its value on the bottom $\varphi|_{z=0} = A(x, y, t)$ as the power series

$$\varphi(x, y, z, t) = \sum_{n=0}^{\infty}(-1)^n \varepsilon^{2n} \frac{z^{2n}}{(2n)!} \Delta_2^n A(x, y, t),$$

where $\Delta_2 = \partial_x^2 + \partial_y^2$ is the Laplace operator with respect to the horizontal variables x and y. By this representation, the boundary conditions on the free surface $z = h(x, y, t)$ lead, in the approximation of lower order with respect to ε, to the *shallow water* equations for the functions h and $\mathbf{u} = \nabla A$ which can be written in dimensional variables as

$$h_t + \text{div}(h\mathbf{u}) = 0,$$

$$\mathbf{u}_t + (\mathbf{u} \cdot \nabla)\mathbf{u} + g\nabla h = 0,$$

(3.8)

where the operations div and ∇ hold with respect to x and y. By the Cauchy–Lagrange integral, the pressure p in the fluid layer is expressed as

$$p(x, y, z, t) = \rho_0 g(h(x, y, t) - z)$$

with accuracy $O(\varepsilon^2)$, i.e., the pressure distribution is hydrostatic.

There is an analogy between gas dynamics and shallow water theory. Denoting by $\rho = h$ the "density" and by $P = \frac{1}{2}gh^2$ the "pressure," we can write (3.8) as

$$\rho_t + \text{div}\,(\rho\mathbf{u}) = 0,$$

$$\mathbf{u}_t + (\mathbf{u} \cdot \nabla)\mathbf{u} + \frac{1}{\rho}\nabla P = 0, \tag{3.9}$$

$$P = \frac{1}{2}g\rho^2,$$

which coincides with the equations of the isentropic flow of a polytropic gas with polytropic exponent $\gamma = 2$.

Problem 3.2 At the time $t = 0$, a dam holding a storage basin of depth h_0 is suddenly destroyed. In the shallow water approximation, find the free surface shape $y = h(x, t)$ for $t > 0$, the velocity u_0 of the water front moving along a dry bottom, and the water discharge q_0 in the dam site.

Solution This problem can be regarded as the Cauchy problem for the one-dimensional shallow water equations

$$h_t + (uh)_x = 0,$$

$$u_t + uu_x + gh_x = 0$$

with discontinuous initial data

$$h(x, 0) = \begin{cases} h_0, & x < 0, \\ 0, & x > 0, \end{cases} \qquad u(x, 0) = 0.$$

A gas-dynamic counterpart of this problem is the problem on the flow into vacuum of a gas initially at rest. The continuous solution for $t > 0$ is represented as a simple centered wave propagating to the left through the fluid at rest with the characteristic velocity $c_0 = \sqrt{gh_0}$ equal to the critical velocity for a given water storage basin. The equations for this simple wave have the form

$$u + 2\sqrt{gh} = 2c_0,$$

$$u - \sqrt{gh} = \frac{x}{t}.$$

The first relation is the condition that the Riemann invariant is constant, and the second relation is the equation of a centered family of rectilinear characteristics. Hence we find the explicit dependence of u and h on x and t in the domain $-c_0 < x/t < 2c_0$:

$$u(x, t) = \frac{2}{3t}(x + c_0 t),$$

$$h(x, t) = \frac{1}{9gt^2}(x - 2c_0 t)^2.$$

Thus, the free surface has the shape of a parabola with vertex $x = 2c_0 t$ travelling to the right with the velocity $u_0 = 2c_0 = 2\sqrt{gh_0}$. The depth and velocity of water in the dam site remain unchanged during the whole time of motion: $h(0, t) = (4/9)h_0$ and $u(0, t) = (1/3)c_0$. Thus, we find the water discharge $q_0 = (4/27)h_0 c_0$, i.e., the water amount flowing out from the dam each second.   □

Answer:

$$h(x, t) = \begin{cases} h_0, & x < -c_0 t, \\ (x - 2c_0 t)^2/(9gt^2), & -c_0 t < x < 2c_0 t, \\ 0, & x > 2c_0 t, \end{cases}$$

$$u_0 = 2c_0, \quad q_0 = (4/27)h_0 c_0 \quad (c_0 = \sqrt{gh_0}).$$

Discontinuous solutions to the hyperbolic shallow water equations describe propagation of bore type waves where the depth and velocity of the fluid change abruptly. To describe such motions, the system (3.8) is taken for conservation laws of mass and depth-averaged horizontal momentum equation which, in the case of the one-dimensional motion, can be written as

$$\partial_t h + \partial_x(uh) = 0,$$

$$\partial_t(uh) + \partial_x(u^2 h + \frac{1}{2}gh^2) = 0. \tag{3.10}$$

The system (3.10) implies the relations on the strong discontinuity

$$D[h] = [uh], \quad D[uh] = [u^2 h + \frac{1}{2}gh^2],$$

where D is the bore velocity.

3.4   Shear Flows of Shallow Water

In the case of two-dimensional flows, the Euler equations (3.2) are written in dimensionless variables as

$$u_t + uu_x + vu_y + p_x = 0,$$

$$\varepsilon^2 (v_t + uv_x + vv_y) + p_y = -1,$$

$$u_x + v_y = 0,$$

where $u(x, y, t)$ is the horizontal velocity, $v(x, y, t)$ is the vertical velocity, and $p(x, y, t)$ is the fluid pressure. In the long wave approximation, the vertical momentum equation implies the hydrostatic law

$$p_y = -1.$$

Using the dynamic condition on the free boundary $y = h(x, t)$ and excluding the pressure from the horizontal momentum equation, we obtain the following system of equations written in dimensional variables:

$$u_x + v_y = 0,$$
$$u_t + uu_x + vu_y + gh_x = 0. \tag{3.11}$$

Integrating the continuity equation with respect to y and using the kinematic condition on the free boundary, we obtain the integro-differential equations

$$h_t + \frac{\partial}{\partial x} \left(\int_0^h u(t, x, y')dy' \right) = 0,$$

$$u_t + uu_x + vu_y + gh_x = 0, \tag{3.12}$$

$$v = - \int_0^y u_x(t, x, y')dy',$$

called the *Benney equations* or shallow water equations for shear flows. They coincide with the classical shallow water equations if the horizontal velocity u is independent of the vertical variable y.

Methods for studying equations of the form (3.12) use a generalization of the theory of nonlinear hyperbolic equations. To write these equations in the form convenient for our purposes, we introduce the new independent variable λ by setting $y = Y(x, \lambda, t)$, where $Y(x, \lambda, t)$ is a solution to the Cauchy problem

$$Y_t + \tilde{u}Y_x = \tilde{v},$$

$$Y|_{t=0} = \lambda h_0(x), \tag{3.13}$$

where $\widetilde{u}(x, \lambda, t) = u(x, Y(x, \lambda, t), t)$, $\widetilde{v}(x, \lambda, t) = v(x, Y(x, \lambda, t), t)$, and the function $h_0(x)$ determines the initial location of the free boundary: $h|_{t=0} = h_0(x)$. This change of variables parametrizes the initial domain $D_0 = \{-\infty < x < +\infty, 0 < y < h_0(x)\}$ in such a way that D_0 is represented as the union of the one-parameter family of curves $\Gamma_\lambda : y = \lambda h_0(x)$, where the value $\lambda = 0$ corresponds to the bottom $y = 0$, and $\lambda = 1$ corresponds to the initial location of the free boundary. Then the evolution of each initial curve Γ_λ is described by the equation $y = Y(x, \lambda, t)$. At each time $t \geq 0$, the value $\lambda = 1$ corresponds to the free boundary $y = h(x, t)$.

By the definition of the functions $\widetilde{u}$ and $\widetilde{v}$, we have

$$\widetilde{u}_t = u_t + u_y Y_t, \quad \widetilde{u}_x = u_x + u_y Y_x, \quad \widetilde{u}_\lambda = u_y Y_\lambda,$$
$$\widetilde{v}_\lambda = -u_x Y_\lambda = -(\widetilde{u}_x Y_\lambda - \widetilde{u}_\lambda Y_x).$$

Consequently, $\widetilde{u}_t + \widetilde{u}\, \widetilde{u}_x + gh_x = 0$. Differentiating (3.13) with respect to λ, we get

$$(Y_\lambda)_t + (\widetilde{u} Y_\lambda)_x = 0.$$

Since

$$h(x, t) = \int_0^1 Y_\lambda(x, \lambda, t) d\lambda,$$

we finally obtain the shallow water equations for shear flows

$$(H)_t + (\widetilde{u} H)_x = 0,$$
$$\widetilde{u}_t + \widetilde{u}\, \widetilde{u}_x + g\left(\int_0^1 H d\lambda \right)_x = 0 \qquad (3.14)$$

for $H(x, \lambda, t) = Y_\lambda(x, \lambda, t)$ and $\widetilde{u}(x, \lambda, t)$. The system (3.14) can be treated as an infinite-dimensional system of equations parametrized by the continuous parameter λ in the interval $0 \leq \lambda \leq 1$. Such systems can be represented in an abstract form [31] as follows. Let B be a Banach space of vector-valued functions $\mathbf{u}(x, \lambda, t)$, and let A be a linear operator acting on such vector-valued functions only with respect to λ: $A\langle \cdot \rangle : B \to B$. We consider the equation

$$\mathbf{u}_t + A(\mathbf{u})\langle \mathbf{u}_x \rangle = 0 \qquad (3.15)$$

for $\mathbf{u}$. The characteristic of the system (3.15) is defined by the equation $dx/dt = c(x, t)$, where c is an eigenvalue of the problem

$$(\mathbf{F}, (A - cI)\langle \mathbf{f} \rangle) = 0.$$

Here, $\mathbf{F}$ is a linear continuous functional in the dual B^* of the space B and $\mathbf{f} \in B$ is a test vector. The action of the proper functional $\mathbf{F}$ on the system (3.15) yields the relation on the characteristic

$$(\mathbf{F}, \mathbf{u}_t + c\mathbf{u}_x) = 0.$$

The system (3.15) is called *hyperbolic* if all eigenvalues c are real and the corresponding proper functionals $\mathbf{F}$ form a complete system. In this case, the relations on characteristics are equivalent to the system (3.15).

3.5   Nonlinear Dispersive Equations

Using higher order approximations with respect to the parameter ε, we can take into account the dispersion properties of nonlinear long water waves. We consider the two-dimensional potential motion in the (x, y)-plane. In this case, it is convenient to represent the velocity field in terms of the stream function: $\mathbf{u} = (\psi_y, -\psi_x)$. The stream function and the velocity potential form a pair of conjugate harmonic functions connected by the Cauchy–Riemann equations

$$\varphi_x = \psi_y,$$
$$\varphi_y = -\psi_x.$$

We write the original equations of fluid motion in terms of the dimensionless stream function $\psi' = \psi/(h_0\sqrt{gh_0})$

$$\varepsilon^2\psi_{xx} + \psi_{yy} = 0 \;\; (0 < y < h), \quad \psi = 0 \;\; (y = 0),$$
$$h_t + (\psi_x + \psi_y h_x)|_{y=h} = 0,$$
$$(\psi_{yt} - \varepsilon^2 h_x\psi_{xt})|_{y=h} + \frac{1}{2}\frac{\partial}{\partial x}\{\varepsilon^2\psi_x^2(x, h, t) + \psi_y^2(x, h, t)\} + h_x = 0,$$

where we omit the primes to simplify the notation. The last equation is obtained by differentiating the Cauchy—Lagrange integral with respect to x at points of the free surfaces (the condition $p(x, h(x, t), t) = $ const is taken into account). Further, we introduce the depth-averaged horizontal fluid velocity

$$u(x, t) = \frac{1}{h(x, t)} \int\limits_0^{h(x,t)} \varphi_x(x, y, t)dy$$

such that $\psi|_{y=h} = uh$ for the stream function ψ. The function ψ admits the following representation in terms of u and h in $\Omega(t)$, similar to the Lagrange series for the potential φ:

$$\psi = uy + \frac{1}{6}\varepsilon^2(h^2y - y^3)u_{xx} + O(\varepsilon^4).$$

Substituting this expression into the boundary condition at $y = h(x, t)$ and leaving quantities of order up to ε^2 we obtain the second-order long wave approximation (called the *Serre–Su–Gardner equations* in the one-dimensional case) written in dimensional variables as

$$h_t + (uh)_x = 0,$$

$$u_t + uu_x + gh_x = \frac{1}{3h}(h^3(u_{xt} + uu_{xx} - u_x^2))_x. \tag{3.16}$$

The terms involving third order derivatives yield the dispersion correction to the classical shallow water equations. In the case of small perturbations $h = h_0 + \zeta$, $u = u_0 + v$ of the uniform flow, the linearized equations (3.16) take the form

$$\zeta_t + u_0\zeta_x + h_0v_x = 0,$$

$$v_t + u_0v_x + g\zeta_x = \frac{1}{3}h_0^2(v_t + u_0v_x)_{xx}.$$

Hence for the elementary wave packets

$$\zeta(x, t) = a\exp\{i(kx - \omega t)\},$$
$$v(x, t) = b\exp\{i(kx - \omega t)\}$$

we obtain the dispersion relation

$$(\omega - u_0k)^2 = \frac{gh_0k^2}{1 + \frac{1}{3}h_0^2k^2}.$$

Comparing with the exact dispersion relation (3.6), we see that the linearization of (3.16) is equivalent to the fractional rational approximation of the function $\xi \tanh \xi$ in the limit of long waves or small depth as $\xi = h_0k \to 0$.

The dispersive terms in (3.16) can be written in other form by using the operator of total differentiation $d_t = \partial_t + u\partial_x$ with the average velocity u. The first equation in (3.16) implies the relation

$$d_t^2h \equiv (\partial_t + u\partial_x)^2h = -h(u_{xt} + uu_{xx} - u_x^2)$$

which in dimensional variables leads to another form of the Serre–Su–Gardner equations

$$h_t + (uh)_x = 0,$$
$$u_t + uu_x + gh_x + \frac{1}{3h}(h^2 d_t^2 h)_x = 0. \tag{3.17}$$

As in the case of the usual shallow water equations, this approximate model can be interpreted as the gas-dynamic equations (3.9), but with the following more complicated equation of state

$$P = \frac{1}{2} g \rho^2 + \frac{1}{3} \rho^2 d_t^2 \rho.$$

For the class of motions described by the functions $h = 1 + \varepsilon^2 \zeta$, $u = \varepsilon^2 v$ in dimensional variables the nonlinear dispersion equations are simpler than (3.16) and (3.17). Such a modelling means that we deal with weakly nonlinear dispersive shallow water waves of small amplitude. In this case, the system (3.16) leads to the approximate equation

$$h_t + (uh)_x = 0,$$
$$u_t + uu_x + gh_x - \frac{1}{3} h_0^2 u_{xxt} = 0 \tag{3.18}$$

and the Serre–Su–Gardner system leads to the equations

$$h_t + (uh)_x = 0,$$
$$u_t + uu_x + gh_x + \frac{1}{3} h_0 h_{xtt} = 0, \tag{3.19}$$

Both systems (3.18) and (3.19) are called the *Boussinesq equations*.

An approximate description of long waves travelling in only one direction, to the left or to the right, is obtained in the independent variables $\tau = \varepsilon^2 t$, $\xi = x - c_0 t$ with $c_0^2 = gh_0$. This stretching transformation of t means that the long time wave evolution is observed in the slow time scale. In this case, the system (3.16), up to a quantity of order $O(\varepsilon^4)$, is reduced to the Korteweg-de Vries equation

$$\zeta_t + c_0 \left(1 + \frac{3}{2h_0} \zeta \right) \zeta_x + \frac{1}{6} c_0 h_0^2 \zeta_{xxx} = 0$$

written in the original dimensional variables for the function $\zeta(x, t)$ interpreted as the elevation of the free surface in a weakly nonlinear long wave on the surface of a fluid of finite depth h_0.

The multi-dimensional counterpart of the Serre–Su–Gardner equations is often called the *Green–Naghdi equations*

$$h_t + \text{div}(h\mathbf{u}) = 0,$$

$$d_t\mathbf{u} + g\nabla h + \frac{1}{3h}\nabla(h^2 d_t^2 h) = 0, \tag{3.20}$$

$$d_t = \partial_t + \mathbf{u}\cdot\nabla,$$

where the operations div and ∇ are taken with respect to the horizontal variable $\mathbf{x} = (x, y)$ and the vector $\mathbf{u}$, as in the one-dimensional case, is interpreted as the depth-averaged horizontal fluid velocity

$$\mathbf{u}(\mathbf{x}, t) = \frac{1}{h(\mathbf{x}, t)}\int\limits_{0}^{h(\mathbf{x},t)} \nabla_{\mathbf{x}}\varphi(\mathbf{x}, z, t)dz.$$

Since the vector structure of the Green–Naghdi equations (3.20) is similar to the structure of the main hydrodynamics equations, similar conservation laws and the first motion integrals hold, in particular, the conservation law of total momentum

$$(h\mathbf{u})_t + \text{div}\,(h\mathbf{u}\otimes\mathbf{u} + \widetilde{p}I) = 0,$$

$$\widetilde{p} = \frac{1}{2}gh^2 + \frac{1}{3}h^2 d_t^2 h. \tag{3.21}$$

From the practical point of view it is not convenient to use the momentum equation in the form (3.21) since the momentum flux contains second order time derivatives of the function h. Therefore, instead of the velocity vector $\mathbf{u}$, we introduce the new sought vector-valued function

$$\mathbf{v} = \mathbf{u} + \frac{1}{3h}\nabla(h^2 d_t h)$$

and write the local momentum equation (the second equation in (3.20)) as

$$d_t\mathbf{v} + \frac{1}{3h^2}d_t h\nabla(h^2 d_t h) - \frac{1}{3h}d_t(\nabla(h^2 d_t h)) + g\nabla h + \frac{1}{3h}\nabla(h^2 d_t^2 h) = 0,$$

where we used the identity $d_t(\nabla f) = \nabla(d_t f) - (\mathbf{u}'_{\mathbf{x}})^T\nabla f$ for any smooth function $f(\mathbf{x}, t) \in C^2$. After simple transformations, we obtain the relation

$$d_t\mathbf{v} + \left(\frac{\partial\mathbf{u}}{\partial\mathbf{x}}\right)^T(\mathbf{v} - \mathbf{u}) + g\nabla h - \frac{1}{2}\nabla(d_t h)^2 = 0$$

which can be written as

$$\mathbf{v}_t + \operatorname{curl} \mathbf{v} \times \mathbf{u} + \nabla \left(\mathbf{v} \cdot \mathbf{u} - \frac{1}{2} |\mathbf{u}|^2 + gh - \frac{1}{2}(d_t h)^2 \right) = 0, \tag{3.22}$$

where the operation curl and the vector product are applied to the three-dimensional vectors $(\mathbf{v}, 0)$ and $(\mathbf{u}, 0)$. A consequence of (3.22) is the Helmholtz equation in the form

$$\boldsymbol{\Omega}_t + \operatorname{curl} (\boldsymbol{\Omega} \times \mathbf{u}) = 0$$

for the *generalized vorticity* vector $\boldsymbol{\Omega} = \operatorname{curl} \mathbf{v}$. In particular, this implies the conservation of the *generalized circulation*

$$\Gamma = \oint_{C(t)} \mathbf{v} \cdot d\mathbf{x}$$

is preserved along any contour $C(t)$ in the horizontal plane $\mathbf{x} = (x, y)$ consisting of the same particles moving with the velocity field $\mathbf{u}$. Another consequence of (3.22) is a counterpart of the Cauchy–Lagrange integral

$$\varphi_t + \mathbf{v} \cdot \mathbf{u} - \frac{1}{2} |\mathbf{u}|^2 + gh - \frac{1}{2}(d_t h)^2 = b(t)$$

for generalized potential flows in the case $\mathbf{v} = \nabla \varphi$.

3.6   Stationary Surface Waves

In the reference frame moving with a travelling wave, the motion is described by the stationary solution, where the sought functions are independent of t. The problem for two-dimensional stationary surface waves is formulated as follows. Find a stream function $\psi(x, y)$ and a function $h(x) > 0$ such that

$$\psi_{xx} + \psi_{yy} = 0 \quad (0 < y < h(x)),$$
$$\psi(x, 0) = 0, \quad \psi(x, h(x)) = Q, \tag{3.23}$$
$$\psi_x^2 + \psi_y^2 + 2gh = 2b \quad (y = h(x)),$$

where b is the Bernoulli constant and Q is the flow discharge, constant on each vertical cross-section of the layer.

The problem (3.23) can be reduced to finding solutions to a single integro-differential equation for a function determining the free surface shape. The complex potential $w(z) = \varphi(x, y) + i\psi(x, y)$ realizes a conformal mapping from the flow domain in the plane $z = x + iy$ to the strip $-\infty < \varphi < +\infty$, $0 < \psi < Q$. Passing to dimensionless variables in which the strip has unit width, we look for the inverse mapping $z = w + Z(w)$. For solitary wave type solutions the analytic function $Z(w) = X(\varphi, \psi) + iY(\varphi, \psi)$ in the strip $0 < \psi < 1$ should satisfy the boundary conditions

$$Y = 0 \quad (\psi = 0),$$

$$\frac{1}{(1 + X_\varphi)^2 + Y_\varphi^2} + 2F^{-2}Y = 1 \quad (\psi = 1),$$

where $F^2 = Q^2/gh_0^3$ is the square Froude number (h_0 is the unperturbed fluid depth). Furthermore, the decay condition $Z \to 0$ as $|\varphi| \to +\infty$ should hold. We introduce the function $\eta(\varphi) = Y(\varphi, 1)$ describing the free boundary shape. We define the auxiliary operator N of "normal derivative" sending $\eta(\varphi)$ to the derivative $Y_\psi(\varphi, 1)$, where Y is the solution to the Dirichlet problem

$$Y_{\varphi\varphi} + Y_{\psi\psi} = 0 \quad (0 < \psi < 1),$$

$$Y(\varphi, 0) = 0, \quad Y(\varphi, 1) = \eta(\varphi).$$

We consider the analytic function $f(w) = 1/(1 + dZ/dw) - 1$, where $\mathrm{Im}\, f = 0$ at $\psi = 0$ and $f(w) \to 0$ as $|\varphi| \to +\infty$. We have

$$\mathrm{Re}\, f = \frac{1 + X_\varphi}{(1 + X_\varphi)^2 + Y_\varphi^2} - 1,$$

$$\mathrm{Im}\, f = -\frac{Y_\varphi}{(1 + X_\varphi)^2 + Y_\varphi^2}.$$

By the Cauchy–Riemann equations $(\mathrm{Re}\, f)_\varphi = (\mathrm{Im}\, f)_\psi$ and the definition of N, we obtain the following relation at the boundary $\psi = 1$:

$$\left(\frac{1 + X_\varphi}{(1 + X_\varphi)^2 + Y_\varphi^2} \right)_\varphi = -N \frac{Y_\varphi}{(1 + X_\varphi)^2 + Y_\varphi^2}.$$

Making simple transformations and taking into account the boundary conditions for the function Z, we obtain the required equation [16] for the function η

$$\eta - F^2\eta + \eta N\eta + \frac{1}{2}N(\eta^2) = 0.$$

The nonzero solution corresponding to a solitary surface wave bifurcates from the zero solution $\eta = 0$ at the point $F = 1$ and exists in the interval $1 < F < 1.290$ (the numerical estimate).

An approximate description of solutions to the problem (3.23) can be obtained by looking for stationary solutions to the Serre–Su–Gardner equations (3.17). In this case, from the first equation in (3.17) we obtain the flow discharge integral $uh = Q = \mathrm{const}$. Excluding u from the second equation with the help of the above integral and integrating twice, we obtain the first order ordinary differential equation (the Boussinesq–Rayleigh equation) for the function h

$$\frac{1}{3}Q^2\left(\frac{dh}{dx}\right)^2 = -gh^3 + bh^2 - 2ch + Q^2, \tag{3.24}$$

where b and c are the integration constants connected with the roots h_i $(i = 1, 2, 3)$ of the cubic polynomial on the right-hand side of the Vieta formulas

$$h_1 + h_2 + h_3 = \frac{2b}{g}, \quad h_1 h_2 h_3 = \frac{Q^2}{g}.$$

The solutions are nontrivial if all the roots are real and the function h takes the values in the interval $h_1 \leqslant h_2 < h < h_3$. The free surface shape is implicitly given by the quadrature formula

$$x = \frac{Q}{\sqrt{3g}} \int_h^{h_3} \frac{ds}{\sqrt{(s - h_1)(s - h_2)(h_3 - s)}}.$$

Consequently, as in the case of the Korteweg-de Vries equation (cf. Sect. 2.6), the shape of the free surface is given by the equation

$$h(x) = h_2 + (h_3 - h_2)cn^2(rx; m),$$

$$r = \frac{\sqrt{3g(h_3 - h_1)}}{2Q}, \quad m^2 = \frac{h_3 - h_2}{h_3 - h_1}.$$

This cnoidal wave has amplitude $a = (h_3 - h_2)/2$ and is periodic with period

$$L = \frac{2Q}{\sqrt{3g}} \int_{h_2}^{h_3} \frac{ds}{\sqrt{(s - h_1)(s - h_2)(h_3 - s)}}.$$

In the limit of waves of small amplitude $a \to 0$, we obtain the value $m = 0$, so that the Jacobi amplitude becomes a linear function $\beta = rx$ and a cnoidal wave takes the form of an elementary wave packet $h(x) = h_0 + a\cos kx$ with the wavenumber $k = 2r$ and the average fluid depth $h_0 = (h_2 + h_3)/2$. Introducing the phase velocity u_0 by the identity $u_0^2 + 2gh_0 = 2b$ (it is legitimate since the parameter b has the sense of the Bernoulli constant), we obtain the dispersion relation

$$F^2 = \frac{1}{1 + \frac{1}{3}h_0^2 k^2},$$

where $F = u_0/\sqrt{gh_0}$ is the Froude number. Thus, in the linear limit, the stationary wave turns out to be subcritical, which agrees with the conclusions of Sect. 3.2.

In the other limit case, where the root $h_2 \to h_1$ becomes double, the elliptic cosine is transformed to an elementary function: $cn(\xi; m) \to 1/\cosh\xi$ as $m \to 1$. In the limit, the period L of the cnoidal wave infinitely increases and is transformed to a *solitary wave* with the amplitude $a = h_3 - h_2 > 0$ and the free surface profile

$$h(x) = h_0 + \frac{a}{\cosh^2 rx}, \qquad r = \frac{\sqrt{3\left(\frac{h_3}{h_2} - 1\right)}}{2\sqrt{\frac{h_3}{h_2}}h_2}.$$

The quantity $h_0 = h_2$ yields the asymptotics of the layer depth as $|x| \to \infty$. Therefore, in this case, it is natural to define the Froude number $F = u_0/\sqrt{gh_0}$ from the fluid velocity u_0 at infinity (i.e., from the condition $u_0^2 + 2gh_2 = 2b$). Hence we obtain the Boussinesq formula for the parameters of the solitary wave

$$F^2 = 1 + \frac{a}{h_0}.$$

By this formula, this nonlinear wave is supercritical.

The constructed solution to the Green–Naghdi equation yields waves of an arbitrarily large amplitude, whereas the exact solution to the problem (3.23) exists only in a finite interval of amplitudes. The limiting solitary or periodic wave of finite amplitude has a cusp with the angle $120°$ at the wave crest.

3.7 Waves in Two-Layer Fluids

Consider the plane potential motion of an ideal fluid consisting of two layers $\Omega_j(t)$ ($j = 1, 2$) of densities $\rho_1 \neq \rho_2$ that are separated by the surface $y = h(x, t)$. We assume that the fluid is bounded by the flat bottom $y = 0$ from below and by an impermeable lid $y = H$ from above.

The velocity potentials φ_j and function h satisfy the equations

$$\varphi_{jxx} + \varphi_{jyy} = 0, \quad \mathbf{x} = (x, y)^T \in \Omega_j(t) \quad (j = 1, 2),$$

$$\varphi_{1y} = 0 \ (y = 0), \quad \varphi_{2y} = 0 \ (y = H),$$

$$\left.\begin{array}{l} h_t + \varphi_{jx}h_x - \varphi_{jy} = 0, \quad (j = 1, 2) \\[4pt] \rho_1\left(\varphi_{1t} + \frac{1}{2}\varphi_{1x}^2 + \frac{1}{2}\varphi_{1y}^2 + gh\right) \\[8pt] = \rho_2\left(\varphi_{2t} + \frac{1}{2}\varphi_{2x}^2 + \frac{1}{2}\varphi_{2y}^2 + gh\right) \end{array}\right\} \quad y = h(x, t).$$

We recall that the kinematic condition admits a nonzero jump of the tangent fluid velocity through the interface, whereas the dynamic condition means the continuity of pressure.

The linear theory deals with small perturbations of the piecewise constant horizontal flow with rectilinear boundary $h(x, t) = h_1 = $ const and constant velocities u_j (i.e., with the potentials $\varphi_{0j}(x, y, t) = u_j x - (gh_1 + \frac{1}{2}u_j^2)t$). Looking for solutions to the linearized equations in the form of wave packets

$$h = h_1 + ae^{i(kx - \omega t)},$$

$$\varphi_1 = \varphi_{01} + A_1 \cosh ky e^{i(kx - \omega t)},$$

$$\varphi_2 = \varphi_{02} + A_2 \cosh k(H - y)e^{i(kx - \omega t)}$$

with constant amplitudes a, A_1, and A_2, we obtain the dispersion relation

$$\rho_1(\omega - u_1 k)^2 \coth kh_1 + \rho_2(\omega - u_2 k)^2 \coth k(H - h_1) = (\rho_1 - \rho_2)gk \qquad (3.25)$$

which is the quadratic equation for the frequency ω. If this equation has a pair of complex roots, then the main flow is unstable. Thus, the frequency ω is complex for all wavenumbers k in the case $\rho_1 < \rho_2$, where a heavier fluid occurs in the upper layer. Such an instability of the layer interface is called the *Rayleigh–Taylor instability*. If $\rho_1 > \rho_2$, then the roots are real if and only if

$$(u_2 - u_1)^2 \leqslant \frac{\mu}{\lambda}(\tanh k(H - h_1) + \lambda \tanh kh_1)\frac{g}{k}, \qquad (3.26)$$

where $\lambda = \rho_2/\rho_1$ and $\mu = 1 - \lambda$. If $u_2 \neq u_1$ (the relative velocity of the motion of layers differs from zero), then the inequality (3.26) fails for sufficiently large $|k|$. Consequently, in the case of sliding of layers, we have the short-wave instability, called the *Kelvin–Helmholtz instability*.

The shallow water approximation for a two-layer fluid is obtained by representing the potentials $\varphi_j(x, y, t)$ as the Lagrange series

$$\varphi_1 = \sum_{n=0}^{\infty}(-1)^n \varepsilon^{2n} \frac{y^{2n}}{(2n)!} \partial_x^{2n} A_1(x, t),$$

$$\varphi_2 = \sum_{n=0}^{\infty}(-1)^n \varepsilon^{2n} \frac{(H - y)^{2n}}{(2n)!} \partial_x^{2n} A_2(x, t),$$

where $A_1(x, t) = \varphi_1|_{y=0}$ and $A_2(x, t) = \varphi_2|_{y=H}$. The two-layer shallow water equations for h, $u_1 = A_{1x}$, and $u_2 = A_{2x}$ have the form

$$h_t + (u_1 h)_x = 0,$$

$$(H - h)_t + (u_2(H - h))_x = 0,$$

$$u_{1t} + u_1 u_{1x} + \mu gh_x = \lambda(u_{2t} + u_2 u_{2x}).$$

Summarizing the first two equations, we get

$$u_1 h + u_2(H - h) = Q(t),$$

where Q is an arbitrary function of t. Assuming that the flow is constant at $x \pm \infty$ and using the Galilean invariance property of the equations of motion, we can assume without loss of generality that $Q = 0$. Then we obtain an expression for the fluid velocity in the upper layer $u_2 = -u_1 h/(H - h)$. As a result, the equations of two-layer shallow water under a rigid lid are reduced to the system of two equations

$$h_t + (u_1 h)_x = 0,$$

$$\frac{H - \mu h}{H - h} u_{1t} + \frac{H^2 - (2 + \lambda)Hh + \mu h^2}{(H - h)^2} u_1 u_{1x} \qquad (3.27)$$

$$+ \left(\mu g - \frac{\lambda H^2}{(H - h)^3} u_1^2\right) h_x = 0.$$

This system of equations is hyperbolic if and only if

$$\frac{u_1^2}{gH} < \frac{\mu}{\lambda}\left(1 - \frac{h}{H}\right)^2\left(1 - \mu\frac{h}{H}\right). \qquad (3.28)$$

It is easy to verify that the inequality (3.28) agrees with the stability criterion (3.26) in the limit of long waves as $k \to 0$. Thus, the two-layer shallow water equations form an evolution system of mixed type unlike the case of one-layer shallow water equations that are hyperbolic equations. A simpler approximate model is obtained by the additional simulation

$$h \to H\frac{1 + A}{2}, \quad u_2 - u_1 \to \sqrt{\mu g H}B, \quad t \to \frac{2}{\sqrt{\mu g H}}t$$

with small parameter μ (approximation in the case of fluid flows with close densities). Considering the terms of leading order with respect to the parameter μ, we obtain the system of equations

$$A_t + ((A^2 - 1)B)_x = 0,$$

$$B_t + ((B^2 - 1)A)_x = 0. \qquad (3.29)$$

Nonlinear stationary waves in a two-layer fluid under a rigid lid are described by the following equations for the stream functions $\psi_j(x, y)$ ($j = 1, 2$) and the function $h(x)$ ($0 < h(x) < H$):

$$
\begin{aligned}
&\psi_{1xx} + \psi_{1yy} = 0 \quad (0 < y < h(x)), \\
&\psi_{2xx} + \psi_{2yy} = 0 \quad (h(x) < y < H), \\
&\psi_1(x, 0) = 0, \quad \psi_1(x, h(x)) = \psi_2(x, h(x)) = Q_1, \\
&\psi_2(x, H) = Q_1 + Q_2, \\
&\rho_1(\psi_{1x}^2 + \psi_{1y}^2 + 2gh - 2b_1) \\
&\qquad = \rho_2(\psi_{2x}^2 + \psi_{2y}^2 + 2gh - 2b_2) \quad (y = h(x)),
\end{aligned}
\tag{3.30}
$$

where Q_j is the water discharge and b_j is the Bernoulli constant for the jth layer. In the absence of waves, the flow with rectilinear interface and constant velocities is described by the solution

$$
\begin{aligned}
&h(x) = h_1 = \text{const} \quad (0 < h_1 < H), \\
&\psi_1(x, y) = u_1 y \quad (0 < y < h_1), \\
&\psi_2(x, y) = u_2(y - h_1) + u_1 h_1 \quad (h_1 < y < H).
\end{aligned}
$$

This solution is obtained for the following values of the discharges and Bernoulli constants:

$$
\begin{aligned}
&Q_1 = h_1 u_1, \quad Q_2 = h_2 u_2, \\
&b_1 = u_1^2 + 2gh_1, \quad b_2 = u_2^2 + 2gh_1,
\end{aligned}
\tag{3.31}
$$

where $h_2 = H - h_1$ is the depth of the unperturbed upper fluid layer. Theses values of Q_j and b_j should be involved in Eqs. (3.30) and in searching solitary wave type solutions with the asymptotic behavior $h(x) \to h_1$, $\nabla \psi_j \to (0, u_j)$ ($j = 1, 2$) as $x \to -\infty$.

The second-order shallow water approximation yields the following nonlinear ordinary differential equation for $h(x)$:

$$
\frac{1}{3}(\rho_1 Q_1^2 (H - h) + \rho_2 Q_2^2 h)\left(\frac{dh}{dx}\right)^2 = P(h),
\tag{3.32}
$$

where $P(h) = h(h - H)((\rho_1 - \rho_2)gh^2 - 2(\rho_1 b_1 - \rho_2 b_2)h + c) + \rho_1 Q_1^2(H - h) + \rho_2 Q_2^2 h$ and c is the integration constant. For solitary waves the conditions $h(x) \to h_1$ and $h'(x) \to 0$ as $x \to -\infty$ imply

$$
c = \rho_1 u_1^2 h_1 + \rho_2 u_2^2 h_2 - g(\rho_1 - \rho_2)h_1^2 + 2(\rho_1 b_1 - \rho_2 b_2)h_1.
$$

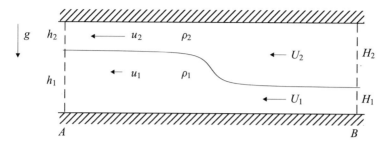

Fig. 3.3 A sketch for smooth bore type solutions between horizontal top and bottom rigid walls

Equation (3.32), obtained by L. V. Ovsyannikov for an approximate description of stationary waves in a two-layer fluid, is a counterpart of the Boussinesq–Rayleigh equations (3.24) in the theory of surface waves. If the polynomial $P(h)$ of the 4th degree has one double root $h = h_1$ and two simple roots different from h, then Eq. (3.32) has a solution in the form of a symmetric solitary wave of elevation or depression. In the case of two double roots, we have nonsymmetric smooth bore type waves (cf. Fig. 3.3).

We write Eq. (3.32) in dimensionless variables by setting $h(x) = h_1(1 + \eta(\bar{x}))$, $x = h_1\bar{x}$ and introducing the ratio $r = h_2/h_1$ and the *densimetric* Froude numbers

$$F_1^2 = \frac{\rho_1 u_{01}^2}{g(\rho_1 - \rho_2)h_1}, \qquad F_2^2 = \frac{\rho_2 u_{02}^2}{g(\rho_1 - \rho_2)h_2}.$$

According to Eq. (3.32), for the function η expressing the deviation of the interface from the equilibrium state we obtain the equation

$$\left(\frac{d\eta}{dx}\right)^2 = \frac{3\eta^2 \left(F_1^2(r - \eta) + rF_2^2(1 + \eta) - (1 + \eta)(r - \eta)\right)}{F_1^2(r - \eta) + r^3 F_2^2(1 + \eta)}. \tag{3.33}$$

The interval $-1 < \eta < r$ corresponds to the domain of admissible values $0 < h(x) < H$. Since for such η the denominator of the fraction on the right-hand side of (3.33) is positive, the numerator should be positive in a neighborhood of $\eta = 0$, which is valid if and only if F_1 and F_2 satisfy the inequality

$$F_1^2 + F_2^2 > 1. \tag{3.34}$$

This condition has a certain sense from the point of view of the dispersion properties of nonlinear waves. Namely, in the case of stationary waves (i.e., wave packets with the frequency $\omega = 0$ and unknown dimensionless wavenumber $\xi = kh_0$), the dispersion relation (3.25) is written in dimensionless variables as

$$F_1^2\xi \coth \xi + F_2^2 r\xi \coth r\xi = 1.$$

Since the even function $f(\xi) = \xi \coth \xi$ is strictly monotonically increasing for $\xi \geqslant 0$, the real roots $\xi \in \mathbb{R}$ can exist only if $F_1^2 + F_2^2 \leqslant 1$. Thus, linear stationary waves in a two-layer fluid of the form of sinusoidal wave packets exist only in the *subcritical* domain $F_1^2 + F_2^2 \leqslant 1$, whereas Eq. (3.33) describes nonlinear waves existing in the *supercritical* domain (3.34). In particular, smooth bore type solutions are obtained for the Froude numbers

$$|F_1| = \frac{1+a}{\sqrt{1+r}}, \quad |F_2| = \frac{r-a}{\sqrt{r(1+r)}},$$

where the dimensionless parameter a $(-1 < a < r)$ is the bore amplitude. The locus of points (F_1, F_2) is the rhombus $|F_1| + \sqrt{r}|F_2| = \sqrt{1+r}$ which touches the inscribed circle $F_1^2 + F_2^2 = 1$ at the points $(\pm 1/\sqrt{(1+r)}, \pm\sqrt{r}/\sqrt{1+r})$ corresponding to the value $a = 0$ (cf. Fig. 3.4). For small a the bore profile is given by the approximate formula

$$\eta(\bar{x}) = \frac{a}{2}\left(1 + \tanh\frac{\kappa a\bar{x}}{2}\right), \quad \kappa^2 = \frac{3}{r(1-r+r^2)}.$$

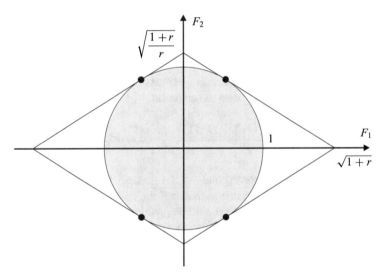

Fig. 3.4 The unit circle in the plane of densimetric Froude numbers (F_1, F_2) separates the subcritical (inside the circle) and supercritical (outside the circle) domains. The rhombus touches the circle at four points from which a one-parameter family of smooth bore type solutions bifurcates.

3.8   Waves in Stratified Fluids

The system (3.1) at rest $\mathbf{u} = 0$ is reduced to the equation

$$\nabla p = \rho \mathbf{g}$$

which implies $p = p_0(z)$, $\rho = \rho_0(z)$, where p_0 and ρ_0 are connected by the hydrostatic law

$$\frac{dp_0}{dz} = -g\rho_0(z).$$

Consequently, under the action of the gravity force, an inhomogeneous fluid at rest splits by the planes $z = \text{const}$ into horizontal layers (strata) of constant density. This phenomenon is called *stratification* and the fluid is referred to as a *stratified fluid*. In this sense, the two-layer fluids considered in the previous section provide the simplest examples of stratification, where the density is piecewise constant. The continuous stratification can also occur in the horizontal *shear flow* described by the exact solution $u = u_0(z)$, $v = v_0(z)$, $w = 0$, $\rho = \rho_0(z)$, $p = p_0(z)$ with arbitrary smooth functions u_0, v_0, ρ_0, and the pressure p_0 connected with ρ_0 by the hydrostatic law.

In the general case, from the second equation in (3.1) it follows that the density ρ is constant along the trajectories of fluid particles, i.e., the integral curves of the system of ordinary differential equations

$$\frac{d\mathbf{x}}{dt} = \mathbf{u}(\mathbf{x}, t).$$

Therefore, each level surface of the density $\rho(\mathbf{x}, t) = \text{const}$ consists of the same fluid particles, which means that it is conserved as a material object. Perturbations propagating along such interfaces within a fluid are called *internal waves*.

To describe the wave motion in a fluid layer of finite depth $0 < z < h(x, y, t)$ bounded by the flat bottom $z = 0$ and the free surface $z = h(x, y, t)$, we impose the boundary conditions

$$w = 0 \quad (z = 0),$$

$$h_t + uh_x + vh_y = w, \quad p = p_0 = \text{const} \quad (z = h).$$

To exclude the surface waves from consideration, the rigid impermeable lid $z = h_0$ is taken for the upper layer boundary, where the impermeability condition $w = 0$ is imposed.

Small perturbations $\mathbf{u} = \mathbf{u}_0 + \mathbf{u}'$, $\rho = \rho_0 + \rho'$, $p = p_0 + p'$ of the shear flow with the velocity vector $\mathbf{u}_0 = (u_0(z), v_0(z), 0)^T$ are described by the linearized system of equations

$$u'_x + v'_y + w'_z = 0,$$

$$D_0\rho' + \rho_{0z}w' = 0,$$

$$D_0 u' + \frac{1}{\rho_0}p'_x + u_{0z}w' = 0,$$

$$D_0 v' + \frac{1}{\rho_0}p'_y + v_{0z}w' = 0,$$

$$D_0 w' + \frac{1}{\rho_0}p'_z + \frac{g}{\rho_0}\rho' = 0,$$

where $D_0 = \partial_t + u_0\partial_x + v_0\partial_y$. Successively excluding the sought functions, we reduce the system to a single equation for the vertical fluid velocity w'

$$\rho_0(N^2 + D_0^2)\Delta_2 w' + D_0^2\frac{\partial}{\partial z}\left(\rho_0\frac{\partial w'}{\partial z}\right)$$

$$- D_0\left((\rho_0 u_{0z})_z\frac{\partial w'}{\partial x} + (\rho_0 v_{0z})_z\frac{\partial w'}{\partial y}\right) = 0, \qquad (3.35)$$

where $\Delta_2 = \partial_x^2 + \partial_y^2$. The quantity $N = N(z)$ is defined by the formula

$$N^2 = -\frac{g\rho_{0z}}{\rho_0}$$

and has the same dimension as the frequency. The quantity $N = N(z)$ is called the *Brunt–Väisälä frequency* (or the *buoyancy frequency*). To describe internal waves in a weakly stratified fluid, one often uses the Boussinesq approximation (do not confuse with the Boussinesq model in the shallow water theory!). In accordance with the Boussinesq approximation, the coefficients ρ_0 and N in (3.35) are constants: $\rho_0(z) \approx \rho_{00} = \mathrm{const} > 0$ and $N(z) \approx N_0 = \mathrm{const} \neq 0$. If there is no velocity shear, we have the equation

$$\Delta w_{tt} + N_0^2\Delta_2 w = 0,$$

where $\Delta = \partial_x^2 + \partial_y^2 + \partial_z^2$.

Example 3.2 Consider the process of generating internal waves in a weakly stratified fluid occupying the whole space $\mathbb{R}^3$. We assume that a point source of perturbations is concentrated at the origin $\mathbf{x} = 0$, oscillates with a small amplitude and a given frequency ω_0, and perturbs the wave field in the form of elementary wave packets $w = a\exp\{i(kx + ly + mz - \omega_0 t)\}$. In this case, the components of the wave vector $\mathbf{k} = (k, l, m)^T$ are connected with the frequency ω_0 by the dispersion relation

$$\omega_0^2 = N_0^2 \frac{k^2 + l^2}{k^2 + l^2 + m^2}.$$

It is clear that the frequency of generated waves cannot exceed the Brunt–Väisälä frequency: $\omega_0 \leq N_0$. Furthermore, the angle β_0 at which the wave vector is inclined to the horizontal Oxy-plane is known:

$$\beta_0 = \text{arc cos} \left(\frac{\omega_0}{N_0} \right).$$

Since the energy of wave packets is transferred from the perturbation source along the direction of the group velocity vector $\mathbf{c}_g = \nabla \omega(\mathbf{k})$, it is of interest to understand the relative orientation of the vectors $\mathbf{c}_g$ and $\mathbf{k}$. By the dispersion relation, the frequency $\omega = \omega(\mathbf{k})$ can be regarded as a homogeneous function of degree 0 in $\mathbf{k}$. Therefore, the group velocity vector is always orthogonal to the wave vector: $\mathbf{k} \cdot \mathbf{c}_g(\mathbf{k}) = 0$. Consequently, the wave motion perturbed by a point source of oscillations is concentrated in a neighborhood of a conical surface with the axis coinciding with the Oz-axis and the opening angle $2\beta_0$.

In the case of a stratified fluid layer of finite deep h_0 that is unbounded in the horizontal direction, the internal waves are described by elementary wave packets

$$w(x, y, z, t) = W(z)e^{i(kx+ly-\omega t)},$$

where the amplitude W depends on the vertical variable z. We denote

$$\tau(z) = ku_0(z) + lv_0(z) - \omega, \quad m^2 = k^2 + l^2.$$

In view of (3.35), for the amplitude W we obtain the differential equation

$$\frac{d}{dz} \left(\rho_0 \tau^2 \frac{d}{dz} \frac{W}{\tau} \right) = \rho_0 m^2 (\tau^2 - N^2) \frac{W}{\tau} \quad (0 < z < h_0).$$

The boundary conditions for a fluid under a rigid lid take the form

$$W(0) = W(h_0) = 0$$

and, in the presence of the free surface,

$$W = 0 \quad (z = 0),$$

$$\tau^2 \frac{d}{dz} \frac{W}{\tau} = gm^2 \frac{W}{\tau} \quad (z = h_0).$$

In these equations with a given wave vector $\mathbf{k} = (k, l)^T$ and the unknown function W, the parameter ω involved in the coefficient τ and the condition on the free surface is also unknown. The values of ω for which there is a nonzero solution W form the *spectrum*. In the case of internal waves propagating in a fluid at rest ($u_0 = v_0 = 0$) under a rigid lid, the spectral problem takes the form

$$(\rho_0 W_z)_z = \rho_0 m^2 \left(1 - \frac{N^2}{\omega^2}\right) W \quad (0 < z < h_0),$$

$$W(0) = W(h_0) = 0.$$

(3.36)

If the upper boundary of the fluid layer is free, then the boundary conditions take the form

$$W(0) = 0, \quad W'(h_0) = g\frac{m^2}{\omega^2}W(h_0).$$

In the general case, the spectrum of the problem (3.36) can be complex. We note that if ω^2 is complex, then the eigenfunction W is also complex. Multiplying by the complex-conjugate function $\overline{W}$, integrating over the interval $(0, h_0)$, and taking into account the boundary conditions, we obtain the relation

$$\omega^2 \int_0^{h_0} \rho_0(|W_z|^2 + m^2|W|^2)dz = g\rho_0(h_0)|W(h_0)|^2 + \int_0^{h_0} \rho_0 N^2|W|^2 dz.$$

If the density $\rho_0(z)$ does not decrease with decreasing z, we have $N^2(z) \geq 0$ for all $z \in (0, h_0)$. In this case, $\omega^2 > 0$ and the spectrum is real. The following assertions give more detailed information on properties of the spectrum of the problem (3.36).

(a) There exists a countable family of wave modes $\omega_n^2(m)$ ($n = 1, 2, \ldots$) such that $\omega_1^2(m) > \omega_2^2(m) > \ldots > \omega_n^2(m) > \ldots$ and $\omega_n^2(m) \to 0$ as $n \to \infty$.
(b) The eigenfunction $W_n(z)$ has exactly n zeros on the interval $[0, h_0]$ if the upper boundary is free and exactly $n + 1$ zeros in the case of the wave motion under a rigid lid.
(c) For each wave mode the phase velocity $c_n(m) = \omega_n(m)/m$ is a strictly monotonically decreasing function of the parameter m (the modulus of the wave vector).

Problem 3.3 In the Boussinesq approximation, find the spectrum of phase velocities and eigenfunctions of the problem on two-dimensional linear internal waves in the layer of a stratified fluid under a rigid lid. What is the group velocity for each wave?

Solution We set $m^2 = k^2$ in (3.36) for the plane motion, where k is the wavenumber. We also set $\rho_0(z) = \text{const}$ and $N(z) = N_0 = \text{const}$ in accordance with the

Boussinesq approximation. In the case $\omega \leqslant N_0$, we denote

$$\lambda^2 = -m^2 \left(1 - \frac{N_0^2}{\omega^2} \right).$$

As a result, we obtain the equations

$$W_{zz} + \lambda^2 W = 0 \quad (0 < z < h_0),$$
$$W(0) = W(h_0) = 0,$$

which imply $\lambda = \lambda_n = \pi n/h_0$, $\varphi_n(z) = \sin \lambda_n z$ $(n = 1, 2, \dots)$. Expressing ω in terms of λ, we find the wave modes $\omega_n^2(k) = N_0^2 k^2/(k^2 + \lambda_n^2)$. Respectively, the phase $c_p^{(n)} = \omega_n(k)/k$ and group $c_g^{(n)} = d\omega_n(k)/dk$ velocities have the form

$$c_p^{(n)}(k) = \pm \frac{N_0}{\sqrt{k^2 + \lambda_n^2}},$$

$$c_g^{(n)}(k) = \pm \frac{N_0 \lambda_n^2}{(k^2 + \lambda_n^2)^{3/2}}.$$

It is clear that $|c_g^{(n)}(k)| < |c_p^{(n)}(k)|$ for all $k \neq 0$, where equalities hold only in the long-wave limit $k = 0$ which provides the maximal value $c_p^{(n)} = c_g^{(n)} = N_0 h_0/(\pi n)$ of the phase and group velocities of the nth mode. In the case $\omega > N_0$, there are no nontrivial solutions to the problem under a lid on the upper boundary of the layer. □

3.9 Stability of Stratified Flows

If the spectrum is such that $\operatorname{Im} \omega \leqslant 0$, then the basic flow with the velocity vector $\mathbf{u}_0 = (u_0(z), v_0(z), 0)^T$ is stable; otherwise, we have instability. In particular, if the eigenfunction $W(z)$ of the problem (3.36) has zero at the point $z = h_*$ in $(0, h_0]$ and $N^2(z) < 0$ for $z \in (0, h_*)$, then for the corresponding eigenvalue we have $\omega^2 < 0$. Consequently, the state at rest or, which is equivalent, the uniform flow with $u_0 = \text{const}$ and $v_0 = \text{const}$ is unstable. For example, a constant flow is unstable if the fluid density $\rho_0(z)$ is a monotonically increasing function on $(0, h_0)$.

In order to analyze the stability of the plane shear flow $\mathbf{u}_0 = (u_0(z), 0, 0)^T$, where the wave packets $w(x, z, t) = W(z)e^{i(kx-\omega t)}$ propagate, it is convenient to take the phase velocity $c = c_p(k) = c_r + ic_i$ for the spectral parameter. The instability takes place in the case $c_i > 0$. Replacing the amplitude function $W(z) = \tau(z)\Psi(z)$, for the flow under a rigid lid we obtain the spectral problem

$$(\rho_0(u_0 - c)^2 \Psi_z)_z + \rho_0(N^2 - k^2(u_0 - c)^2)\Psi = 0 \quad (0 < z < h_0),$$
$$\Psi(0) = \Psi(h_0) = 0. \tag{3.37}$$

Then the dimensionless *Richardson number*

$$Ri(z) = \frac{N^2(z)}{(du_0(z)/dz)^2}$$

is responsible for the flow stability.

Theorem 3.1 [26] *If $Ri(z) \geq 1/4$ for all $z \in (0, h_0)$, then the plane shear flow under a rigid lid is stable.*

Proof Introduce the function $G = (u_0 - c)^{1/2}\Psi$. Then the equation in (3.37) takes the form

$$(\rho_0(u_0 - c)G_z)_z - \frac{1}{2}(\rho_0 u_{0z})_z G - k^2\rho_0(u_0 - c)G + \rho_0\frac{N^2 - \frac{1}{4}u_{0z}^2}{u_0 - c}G = 0.$$

Multiplying this equation by the complex-conjugate function $\overline{G}$, integrating over the interval $(0, h_0)$, taking into account the homogeneous boundary conditions for G, and separating the imaginary part, we obtain the relation

$$c_i\left\{\int_0^{h_0}\rho_0(|G_z|^2 + k^2|G|^2)dz + \int_0^{h_0}\rho_0\left(N^2 - \frac{1}{4}u_{0z}^2\right)\left|\frac{G}{u_0 - c}\right|^2 dz\right\} = 0.$$

Thus, if $Ri(z) \geq 1/4$ for all $z \in (0, h_0)$, then $c_i = 0$.   □

The Howard semicircle theorem specifies the location of the spectrum of phase velocities in the unstable case.

Theorem 3.2 [21] *If $N^2(z) \geq 0$ for $z \in (0, h_0)$, then the complex phase velocities of all unstable modes are located in the semicircle of diameter $(b - a)/2$ and center $((a + b)/2, 0)$ at the real axis, where*

$$a = \min_{z \in [0, h_0]} u_0(z), \quad b = \max_{z \in [0, h_0]} u_0(z).$$

Proof Multiplying the equation in (3.37) by $\overline{\Psi}$, integrating over $(0, h_0)$, taking into account the boundary conditions for Ψ, and separating the real and imaginary parts, we obtain two integral relations

$$\int_0^{h_0}(u_0^2 - 2c_r u_0 + c_r^2 - c_i^2)Q\,dz = B,$$

$$\int_0^{h_0}u_0 Q\,dz = c_r A,$$

where

$$Q = |\Psi_z|^2 + k^2|\Psi|^2, \quad A = \int_0^{h_0} Q \, dz, \quad B = \int_0^{h_0} \rho_0 N^2 |\Psi|^2 dz.$$

From these integral relations we find

$$\int_0^{h_0} (u_0 - a)(u_0 - b)dz = (c_r^2 - (a+b)c_r + c_i^2 + ab)A + B.$$

Since $A > 0$, $B \geqslant 0$, and the integral on the left-hand side is not positive, we obtain the inequality

$$\left(c_r - \frac{a+b}{2}\right)^2 + c_i^2 \leqslant \left(\frac{a-b}{2}\right)^2.$$

The theorem is proved. □

3.10 Stationary Internal Waves

We consider two-dimensional internal waves propagating along the x-axis with constant velocity u_0 in a stratified fluid layer located between the flat bottom $y = 0$ and the rigid lid $y = h_0$. The flow is stationary in the reference frame moving with the wave. In the stationary case, the equations of motion (3.1) take the form

$$u_x + v_y = 0,$$
$$u\rho_x + v\rho_y = 0,$$
$$\rho(uu_x + vu_y) + p_x = 0,$$
$$\rho(uv_x + vv_y) + p_y = -\rho g.$$

We introduce the stream function ψ by the equalities $u = \psi_y$, $v = -\psi_x$. Then the continuity equation div $\mathbf{u} = 0$ is satisfied identically and the equation for the density $\mathbf{u} \cdot \nabla\rho = 0$ can be integrated, which yields the dependence $\rho = \rho(\psi)$. In view of this property, the momentum equations admit the Bernoulli integral written in terms of the stream function as

$$\frac{1}{2}|\nabla\psi|^2 + \frac{1}{\rho(\psi)}p + gy = b(\psi),$$

where $b(\psi)$ is the Bernoulli function. Eliminating the pressure p from the equations of motion and the above integrals, we obtain the second order quasilinear elliptic equation for the stream function

$$\rho(\psi)\Delta\psi + \frac{d\rho(\psi)}{d\psi}\left(gy + \frac{1}{2}|\nabla\psi|^2\right) = \frac{dH(\psi)}{d\psi}, \tag{3.38}$$

where $H(\psi) = \rho(\psi)b(\psi)$. Equation (3.38) is called the *Dubreil-Jacotin–Long equation*. It is possible to specify $\rho(\psi)$ and $b(\psi)$ for flows of solitary wave type by conditions at infinity. In particular, if the flow converges to the uniform flow with the stream function $\psi_\infty(y) = u_0 y$ and known density profile $\rho_\infty(y)$ as $|x| \to \infty$, then

$$\rho(\psi) = \rho_\infty(\psi/u_0),$$

$$\frac{dH(\psi)}{d\psi} = \frac{d\rho(\psi)}{d\psi}\left(\frac{g\psi}{u_0} + \frac{1}{2}u_0^2\right).$$

In this case, the boundary conditions on the bottom and lid take the form

$$\psi\big|_{y=0} = 0, \quad \psi\big|_{y=h_0} = u_0 h_0. \tag{3.39}$$

Example 3.3 Let a fluid in an unperturbed state has the exponential profile of density

$$\rho_\infty(y) = \rho_0 \exp(-N^2 y/g) \quad (0 < y < h_0),$$

where the constant $N > 0$ is the Brunt–Väisälä frequency. In this case, the Dubreil-Jacotin–Long equation (3.38) takes the form

$$\Delta\psi + \frac{N^2}{u_0^2}(\psi - u_0 y) = \frac{N^2}{2gu_0}(|\nabla\psi|^2 - u_0^2).$$

With a uniform flow we associate the exact solution $\psi = u_0 y$ to this equation. We introduce the dimensionless Boussinesq parameter σ and the (densimetric) Froude number F by the equalities

$$\sigma = \frac{N^2 h_0}{g}, \quad F = \frac{u_0}{\sqrt{\sigma g h_0}}.$$

The parameter σ characterizes the vertical gradient density of the fluid. It is small in the case of weak stratification. For small σ the finite values of the Froude number F correspond to small velocities u_0, which is a typical situation for internal waves. To construct the long-wave approximation, we proceed with dimensionless variables $(\bar{x}, \bar{y}) = (\sqrt{\sigma}x/h_0, y/h_0)$, $\bar{\psi} = \psi/(u_0 h_0)$, in which the equation is written as

$$\sigma\psi_{xx} + \psi_{yy} + F^{-2}(\psi - y) = \frac{1}{2}\sigma(\sigma\psi_x^2 + \psi_y^2 - 1),$$

$$\psi(x, 0) = 0, \quad \psi(x, 1) = 1$$

(the bar is omitted). This problem is a nonlinear problem on eigenvalues with the spectral parameter $\lambda = F^{-2}$. A solution to this problem is looked for as power series

$$\psi = y + \psi_0 + \sigma\psi_1 + \sigma^2\psi_2 + \ldots,$$

$$\lambda = \lambda_0 + \sigma\lambda_1 + \sigma^2\lambda_2 + \ldots,$$

whose coefficients should satisfy the equations

$$\psi_{0yy} + \lambda_0\psi_0 = 0,$$

$$\psi_{1yy} + \lambda_0\psi_1 = f_1(\psi_0),$$

$$\ldots\ldots\ldots\ldots\ldots\ldots\ldots$$

with the homogeneous boundary conditions

$$\psi_j(x, 0) = \psi_j(x, 1) = 0 \quad (j = 0, 1, \ldots)$$

and right-hand side $f_1 = -\psi_{0xx} + \psi_{0y} + \psi_{0y}^2/2 - \lambda_1\psi_0$. Hence for the internal waves of the principal mode we have $\psi_0(x, y) = a(x) \sin \pi y$, $\lambda_0 = \pi^2$, where the amplitude factor $a(x)$ remains still undefinite. The equation for the amplitude factor is obtained from the orthogonality conditions

$$\int_0^1 f_1(\psi_0) \sin \pi y \, dy = 0$$

which implies the nonlinear ordinary differential equation

$$\frac{d^2a}{dx^2} = \frac{2\pi}{3}a^2 - \lambda_1 a.$$

A solitary wave type solution satisfying the condition $a(x) \to 0$ as $|x| \to \infty$ is obtained for $\lambda_1 < 0$ and has the form

$$a(x) = \frac{9\lambda_1}{4\pi \cosh^2 \frac{\sqrt{|\lambda_1|}x}{2}}.$$

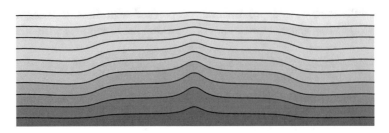

Fig. 3.5 Streamline patterns representing solitary internal waves can be quite complicated. They essentially depend on the density profile at infinity

By the definition of the Froude number F, the velocity of this wave u_0 is connected with the amplitude parameter λ_1 in the approximate formula $u_0^2 = \sigma g h_0/(\pi^2 + \sigma\lambda_1)$. The obtained solution is similar to a solitary wave type solution to the Korteweg-de Vries equation (2.13) and the Boussinesq–Rayleigh equation (3.24). In the general case, the form of the equation for the amplitude function $a(x)$ essentially depends on the density profile $\rho_\infty(y)$ which can cause more complicated forms of solitary internal waves (cf. Fig. 3.5).

3.11   Problems

1. A float rises and falls with a wave 15 times per minute. Find the wavelength L and the wave velocity c if the wave amplitude is small and the fluid depth is infinitely large.
   Answer: $L = 24.98$ m, $c = 6.25$ m/sec.
2. Consider a plane travelling wave in a fluid of finite depth with the free surface $y = h_0 + a\cos(kx - \omega t)$ ($|a| < h_0$ and the parameter ω is connected with k by the dispersion relation $\omega^2 = gk\tanh kh_0$). Find the particle trajectories $\mathbf{x} = \mathbf{x}(t)$ by looking for an approximate solution $\mathbf{x}(t) = (x(t), y(t))$ to the equations

$$\frac{dx}{dt} = \frac{gak}{\omega}\frac{\cosh ky}{\cosh kh_0}\cos(kx - \omega t),$$

$$\frac{dy}{dt} = \frac{gak}{\omega}\frac{\sinh ky}{\cosh kh_0}\sin(kx - \omega t),\qquad(3.40)$$

$$(x, y)|_{t=0} = (\xi, \eta)$$

in the form

$$\mathbf{x}(t) = \mathbf{x}_0(t) + \alpha\mathbf{x}_1(t) + O(\alpha^2),$$

where $\alpha = ak$ is a small parameter. Along what curve in the (x, y)-plane does the point $\mathbf{x}_*(t) = \mathbf{x}_0(t) + \alpha\mathbf{x}_1(t)$ move? Is the motion of the point periodic in t?

Answer:

$$x(t) = \xi + a\frac{\cosh k\eta}{\sinh kh_0}(\sin k\xi - \sin(k\xi - \omega t)) + O(a^2k^2),$$

$$y(t) = \eta - a\frac{\sinh k\eta}{\sinh kh_0}(\cos k\xi - \cos(k\xi - \omega t)) + O(a^2k^2).$$

The point $\mathbf{x}_*(t)$ rotates with period $T = 2\pi/\omega$ along an ellipse.

3. Find an exact solution $\mathbf{x}(t) = (x(t), 0)$ to the system of equations (3.40) describing the horizontal motion of a particle at the bottom $y = 0$ with the initial condition $\mathbf{x}(0) = (0, 0)$ at $t = 0$. Compute the coefficients of the power expansion of this solution

$$x(t) = x_0(t) + ax_1(t) + a^2x_2(t) + O(a^3)$$

with respect to the amplitude a.

Answer:

$$x(t) = \frac{1}{k}(\omega t + f^{-1}(\omega t)),$$

where

$$f(z) = \int_0^z \frac{ds}{\beta\cos s - 1} \quad \left(\beta = \frac{ak}{\sinh kh_0}\right),$$

$$x_0(t) \equiv 0,$$

$$x_1(t) = \frac{\sin \omega t}{\sinh kh_0},$$

$$x_2(t) = \frac{1}{4}k\frac{2\omega t - \sin 2\omega t}{\sinh^2 kh_0}.$$

4. In the linear approximation, consider the three-dimensional wave motion of an infinitely deep fluid with a real velocity potential $\varphi = \text{Im}(\varphi_1 + \varphi_2)$, where φ_j are complex wave packets

$$\varphi_j(\mathbf{x}, z, t) = be^{mz+i(\mathbf{k}_j\cdot\mathbf{x}-\omega t)}, \quad \mathbf{x} = (x, y)^T$$

with different wave vectors $\mathbf{k}_1 = (k, l)^T$, $\mathbf{k}_2 = (-k, l)^T$, but the same frequency $\omega = \sqrt{gm}$ ($m = |\mathbf{k}_1| = |\mathbf{k}_2|$) and amplitude $b \in \mathbb{R}$. Show that the potential φ satisfies the impermeability condition $\varphi_x = 0$ on the vertical wall $x = 0$ and find the shape of the contact line of the free surface $z = \zeta(x, y, t)$ with this wall at each time moment.

Answer:

$$\zeta(0, y, t) = \frac{2b\omega}{g} \cos(ly - \omega t).$$

5. Within the framework of the linear wave theory, find frequencies of eigenoscillations of a fluid with the velocity potential $\varphi = e^{i\omega t} Y(x, y, z)$ in a rectangular reservoir of height h_0, length a, and width b. What is the minimum frequency?
   Answer:

$$\omega_{nm}^2 = g\lambda_{nm} \tanh \lambda_{nm} h_0,$$

$$\lambda_{nm}^2 = \pi^2((n/a)^2 + (m/b)^2),$$

where $n, m = 0, 1, 2, \ldots, n + m \neq 0$;

$$\min_{nm} \omega_{nm}^2 = \frac{\pi g}{a} \tanh \frac{\pi h_0}{a}.$$

6. Consider the linear Cauchy–Poisson problem describing two-dimensional waves on the surface of an infinitely deep fluid (small perturbations of the state at rest)

$$\Phi_{xx} + \Phi_{yy} = 0 \quad (-\infty < y < 0),$$

$$\nabla\Phi \to 0 \quad (y \to -\infty),$$

$$\zeta_t = \Phi_y, \quad \Phi_t + g\zeta = 0 \quad (y = 0),$$

$$\Phi(x, y, 0) = \Phi_0(x, y), \quad \zeta(x, 0) = \zeta_0(x).$$

Show that the function ζ describing the shape of the free surface $y = \zeta(x, t)$ satisfies the integro-differential equation

$$\zeta_{tt}(x, t) + \frac{g}{\pi} v.p. \int_{-\infty}^{+\infty} \frac{\zeta_x(x', t)}{x - x'} dx' = 0$$

with the initial condition

$$\zeta(x, 0) = \eta_0(x), \quad \zeta_t(x, 0) = \Phi_{0y}(x, 0).$$

Hint. Use the Cauchy integral formula for the complex-valued function $f(z, t) = \Phi_{xt} - i\Phi_{yt}$ which is analytic with respect to $z = x + iy$ in the half-plane $\operatorname{Im} z < 0$.

7. Under the conditions of the previous problem, construct the solution $\zeta(x, t)$ with the initial data

$$\zeta(x, 0) = 0, \quad \zeta_t(x, 0) = \frac{2a}{x^2 + a^2} \quad (a > 0 - \text{const}).$$

For this purpose represent ζ as the Fourier integral.
Answer: $\zeta = f_+(x, t) + f_-(x, t)$, where

$$f_\pm(x, t) = \int\limits_0^{+\infty} e^{-ak} \sin(kx \pm \sqrt{gk}t) \frac{dk}{\sqrt{gk}}.$$

8. The frequency ω in a group of waves coming from a distant storm area varied as a linear function and rose by Δ for the observation time period τ. Using the dispersion relation for deep water waves, find the distance r from the storm area.

Answer: $r = \dfrac{g\tau}{2\Delta}$.

9. Show that the phase and group velocities of linear surface waves propagating through a fluid at rest of depth h_0 satisfy the equality

$$c_g(k) = \frac{1}{2} c_p(k) \left(1 + \frac{2kh_0}{\sinh 2kh_0}\right).$$

Using this equality, show that the ratio $c_g(k)/c_p(k)$ is a strictly monotonically decreasing function of the wavenumber k on the half-axis $(0, +\infty)$. What is this ratio in the long ($k \to 0$) and short ($k \to +\infty$) wave limit?
Answer:

$$\lim_{k \to 0} \frac{c_g(k)}{c_p(k)} = 1, \quad \lim_{k \to +\infty} \frac{c_g(k)}{c_p(k)} = \frac{1}{2}.$$

10. Derive the dispersion relation for the plane problem of the linear wave propagation in water of finite depth with the surface tension forces taken into account. The boundary condition for the pressure p on the free surface $y = h_0 + \zeta(x, t)$ in the original nonlinear equations has the form $p = \sigma K$, where $\sigma > 0$ is the surface tension coefficient and K is the curvature of the free surface, expressed by the equality $K = -\zeta_{xx}(1 + \zeta_x^2)^{-3/2}$.
Answer:

$$\omega^2 = gk\left(1 + \frac{\sigma}{\rho g}k^2\right) \tanh kh_0.$$

11. Consider linear surface waves in an infinitely deep fluid with the surface tension forces taken into account.

a. Find the value of the wavenumber $k = k_p$ at which the phase velocity $c_p(k)$ has a minimum.

b. Find the value of the wavenumber $k = k_g$ at which the group velocity $c_g(k)$ has a minimum.

c. Find the value of the wavenumber $k = k_r$ at which the equality $c_p = c_g$ holds (the resonance of phase and group velocities).

d. Find relations between the group velocity c_g and the phase velocity c_p in the short wave limit ($|k| \to +\infty$).

Answer:

a. $k_p = \sqrt{g\rho/\sigma}$.

b. $k_g = k_p \sqrt{\dfrac{2}{\sqrt{3}} - 1}$.

c. $k_r = k_p$.

d. $c_g = (3/2)c_p$.

12. Consider the dispersion relation for three-dimensional linear stationary waves on the surface of an infinitely deep fluid

$$u_0 k + \sqrt{gm} = 0,$$

where u_0 is the velocity of the fluid flow and $m = \sqrt{k^2 + l^2}$ is the modulus of the wave vector $\mathbf{k} = (k, l)^T$ (this relation is obtained from (3.6) in the case of stationary waves ($\omega = 0$) as $h_0 \to \infty$). The wave phase is equal to $\theta = kx + ly$.

a. Passing to the polar coordinates in the plane of the wave vector $\mathbf{k} = m(\cos\alpha, \sin\alpha)^T$ and using the dispersion relation, find the dependence $m = m(\alpha)$.

b. Find the ratio y/x as a function of the polar angle α for which the phase is stationary: $d\theta/d\alpha = 0$.

c. Find the parameter representation $x = x(\alpha), y = y(\alpha)$ of the curve $L(\theta_0)$ along which the wave phase is simultaneously constant $\theta(x, y) = \theta_0 = $ const and stationary. Construct this curve.

d. Find the locus of cusps of the curve $L(\theta_0)$ for different values of the phase θ_0.

Remark   The family of curves $L(\theta_0)$ yields a three-dimensional pattern of waves behind the point source moving with the constant velocity u_0 (ship waves).

Answer:

a. $m(\alpha) = \dfrac{g}{u_0^2 \cos^2\alpha}$.

b. $\dfrac{y}{x} = -\dfrac{\sin\alpha \cos\alpha,}{1 + \sin^2\alpha}$.

c. $x = \dfrac{u_0^2 \theta_0}{g} \cos \alpha (1 + \sin^2 \alpha)$, $y = -\dfrac{u_0^2 \theta_0}{g} \cos^2 \alpha \sin \alpha$.

d. $y = \pm \dfrac{x}{2\sqrt{2}}$.

13. Construct self-similar solutions to the shallow water equations

$$h_t + (uh)_x = 0,$$
$$u_t + uu_x + gh_x = 0$$

that are continuous for $t > 0$ and depend on x/t.

Answer:

$$u(x, t) = \frac{2x}{3t} + C, \quad h(x, t) = \frac{1}{g}\left(\frac{x}{3t} - C\right)^2, \quad C = \text{const} .$$

14. For the shallow water equations find all conservation laws $P_t + Q_x = 0$ with polynomial densities $P(u, h)$ of degree at most 3.

Answer:

$$P(u, h) = C_1 u + C_2 h + C_3 uh + C_4 \left(\frac{1}{2}u^2 h + \frac{1}{2}gh^2\right),$$

$$Q(u, h) = C_1 \left(\frac{1}{2}u^2 + gh\right) + C_2 uh$$

$$+ C_3 \left(u^2 h + \frac{1}{2}gh^2\right) + C_4 \left(\frac{1}{2}u^3 h + guh^2\right).$$

15. Derive the shock adiabat equation in the (u, h)-plane with the origin (u_0, h_0) for a strong discontinuity described by the laws of conservation of mass and total momentum (3.10) of the shallow water equations.

Answer:

$$(u - u_0)^2 = \frac{g(h + h_0)}{2hh_0}(h - h_0)^2.$$

16. Under the conditions of the previous problem, find the expressions for the fluid relative velocities $v = u - D$ and $v_0 = u_0 - D$ on the different sides of the strong discontinuity (D is the discontinuity velocity) in terms of the depths h and h_0 ($h > h_0$). Find out whether these velocities are subcritical or supercritical.

Answer:

$$v = \pm\sqrt{\frac{gh_0(h + h_0)}{2h}}, \quad v_0 = \pm\sqrt{\frac{gh(h + h_0)}{2h_0}},$$

$$|v| < \sqrt{gh}, \quad |v_0| > \sqrt{gh_0}.$$

17. Consider the hydraulic jump (a stationary strong discontinuity for the shallow water equations with the velocity $D = 0$) at which the laws of conservation of mass and total momentum are valid. Show that the energy flux

$$Q = \frac{1}{2}u^3h + guh^2$$

has the jump

$$[Q] = -\frac{gm}{4hh_0}[h]^3$$

with the mass flow $m = uh = u_0h_0$ and the fluid depths h and h_0 satisfying the relation

$$h = \frac{1}{2}h_0(\sqrt{1 + 8F^2} - 1),$$

where $F = u_0/\sqrt{gh_0}$ is the Froude number.

18. Owing to the upstream propagation of a tidal bore along a river, the water level rises by 10% and the flow velocity u_0 is halved. Compute the bore velocity.
    Answer: $D = \frac{9}{2}u_0$.

19. Two fluids layers of depth h_0 for $x < 0$ and h_1 for $x > 0$ ($h_0 > h_1$) are separated by a rigid wall at $x = 0$. At the time $t = 0$, the wall is removed and, as a result, the fluid begins to move. In the shallow water approximation, find the flow velocity $u(x, t)$ and the free surface shape $h(x, t)$ for $t > 0$ in the class of self-similar solutions.
    Answer:

$$u = 0, \quad h = h_0 \quad \left(-\infty < \frac{x}{t} < -\sqrt{gh_0}\right),$$

$$u = \frac{2}{3}\left(\frac{x}{t} + \sqrt{gh_0}\right), \quad h = \frac{1}{9g}\left(\frac{x}{t} - 2\sqrt{gh_0}\right)^2$$

$$\left(-\sqrt{gh_0} < \frac{x}{t} < u_2 - \sqrt{gh_2}\right),$$

$$u = u_2, \quad h = h_2 \quad \left(u_2 - \sqrt{gh_2} < \frac{x}{t} < D\right),$$

$$u = 0, \quad h = h_1 \quad \left(D < \frac{x}{t} < +\infty\right),$$

where $D = u_2h_2/(h_2 - h_1)$ and the constants $u_2 > 0$ and $h_2 > 0$ are found from the relations

$$u_2 + 2\sqrt{gh_2} = 2\sqrt{gh_0},$$

$$u_2^2 = \frac{g(h_1 + h_2)}{2h_1h_2}(h_2 - h_1)^2.$$

20. Show that the linearization of the Benney equations (3.12) at the constant
solution $u = 0$, $v = 0$, $h = h_0 = $ const leads to the equations

$$\frac{\partial h}{\partial t} + \frac{\partial}{\partial x}\left(\int_0^{h_0} u(x, y', t)dy'\right) = 0,$$

$$\frac{\partial u}{\partial t} + g\frac{\partial h}{\partial x} = 0,$$

$$v = -\int_0^y u_x(x, y', t)dy'.$$

21. Show that the linearization of the Benney equations (cf. the previous problem)
admits the Riemann invariants

$$R^\pm = \int_0^{h_0} u(x, y', t)dy' \pm \sqrt{gh_0}h,$$

$$\left(\frac{\partial}{\partial t} \pm \sqrt{gh_0}\frac{\partial}{\partial x}\right)R^\pm = 0,$$

$$\frac{\partial}{\partial t}(u_y) = 0,$$

$$\frac{\partial}{\partial t}\left(u - \frac{1}{h_0}\int_0^{h_0} u(x, y', t)dy'\right) = 0.$$

Give an interpretation of these invariants.

22. Show that the Benney equations (3.12) admit the Riemann invariants

$$R_t^i + c_i R_x^i = 0, \quad i = 1, 2,$$

where

$$R^i(x, t) = c_i - g\int_0^h \frac{dy'}{u(x, y', t) - c_i},$$

and c_i are the roots of the equation

$$1 - g\int_0^h \frac{dy'}{(u(x, y', t) - c_i)^2} = 0,$$

satisfying the inequalities

$$c_1(x, t) < \min_{0 \leqslant y \leqslant h} u(x, y, t) \leqslant \max_{0 \leqslant y \leqslant h} u(x, y, t) < c_2(x, t).$$

23. Verify that the velocity curl $\omega = u_y$ in the long-wave approximation and the invariant

$$R = u - g \int\limits_0^h \frac{dy'}{u(x, y', t) - u(x, y, t)} \tag{3.41}$$

are conserved along the trajectories $dx/dt = u(x, y, t)$, $dy/dt = v(x, y, t)$:

$$\omega_t + u\omega_x + v\omega_y = 0,$$

$$R_t + uR_x + vR_y = 0.$$

The integral in (3.41) is understood in the Cauchy principal value sense.

24. Show that the Benney equations (3.12) have a class of exact solutions with constant vorticity

$$u(x, y, t) = U(x, t) + \Omega y, \quad \Omega = \text{const},$$

if the functions $U(x, t)$ and $h(x, t)$ satisfy the system of equations

$$\frac{\partial h}{\partial t} + \frac{\partial}{\partial x}\left(Uh + \Omega\frac{h^2}{2}\right) = 0,$$

$$\frac{\partial U}{\partial t} + U\frac{\partial U}{\partial x} + g\frac{\partial h}{\partial x} = 0. \tag{3.42}$$

Show that the system (3.42) is hyperbolic. Find Riemann invariants of this system.

25. Under the conditions of the previous problem, find an explicit form of the Riemann invariant (3.41) for the Benney equations (3.12).

Answer:

$$R = u - \frac{g}{\Omega} \ln\left(\frac{h - y}{y}\right).$$

26. For solutions to the system of equations (3.42) we introduce the depth-averaged fluid velocity $W = U + \Omega\frac{h}{2}$. Show that the equations for the functions W and h are written in the conservative form

$$\frac{\partial h}{\partial t} + \frac{\partial}{\partial x}(hW) = 0,$$

$$\frac{\partial hW}{\partial t} + \frac{\partial}{\partial x}(hW^2 + P) = 0,$$

$$P = \frac{gh^2}{2} + \frac{\Omega^2 h^3}{12}.$$

This system is similar to the equations governing the motion of a barotropic, but not polytropic gas. Show that the specific internal energy of such a gas has the form

$$e = \frac{gh}{2} + \frac{\Omega^2 h^2}{24}.$$

27. Verify that the nonlinear kinematic condition for the velocity potential φ on the free boundary $z = h(x, y, t)$ in the three-dimensional Cauchy–Poisson problem is equivalent to the conservative equation

$$h_t + \text{div}_{(x,y)} \int_0^h \nabla_{(x,y)} \varphi(x, y, z, t) dz = 0.$$

28. Consider the two-dimensional Cauchy–Poisson problem. Show that, the tangent velocity $u(x, t) = (\varphi_x + h_x \varphi_y)_{y=h}$ and the normal velocity $v(x, t) = (\varphi_y - h_x \varphi_x)_{y=h}$ of the fluid particles on the free boundary $y = h(x, t)$ satisfy the system of equations

$$h_t = v,$$

$$u_t + \frac{1}{2}\frac{\partial}{\partial x}\left(\frac{u^2 - 2h_x u v - v^2}{1 + h_x^2}\right) + gh_x = 0.$$

29. Show that the two-dimensional Cauchy–Poisson problem admits the integro-differential laws of conservation

$$\frac{\partial}{\partial t}\left(\int_0^h \varphi_x \, dy\right) + \frac{\partial}{\partial x}\left(\int_0^h \left\{\frac{1}{2}\varphi_x^2 - \frac{1}{2}\varphi_x^2 - \varphi_t\right\} dy - \frac{1}{2}gh^2\right) = 0,$$

$$\frac{\partial}{\partial t}\left(\frac{1}{2}\int_0^h (\varphi_x^2 + \varphi_y^2) dy + \frac{1}{2}gh^2\right) - \frac{\partial}{\partial x}\left(\int_0^h \varphi_x \varphi_t \, dy\right) = 0,$$

where the first equality is the depth-averaged differential law of conservation of horizontal momentum, whereas the second one is the law of conservation of energy.

30. Show that the system of equations (3.23) describing plane nonlinear stationary surface waves has the first integral

$$\int_0^{h(x)} (\psi_x^2(x, y) - \psi_y^2(x, y))dy + gh^2(x) - 2bh(x) = c \quad (c = \text{const}).$$

31. Show that the system of equations (3.30) describing plane nonlinear stationary waves in a two-layer fluid under a rigid lid has the first integral

$$\rho_1 \int_0^{h(x)} (\psi_{1x}^2(x, y) - \psi_{1y}^2(x, y))dy + \rho_2 \int_{h(x)}^H (\psi_{2x}^2(x, y) - \psi_{2y}^2(x, y))dy$$

$$+ g(\rho_1 - \rho_2)h^2(x) - 2(\rho_1 b_1 - \rho_2 b_2)h(x) = c \quad (c = \text{const}).$$

32. Using the method of separation of variables, find eigenfunctions of the two-dimensional problem describing stationary surface waves linearized on the supercritical flows with constant depth h_0 and velocity $u_0 > \sqrt{gh_0}$:

$$\Phi_{xx} + \Phi_{yy} = 0 \quad (0 < y < h_0),$$
$$\Phi_y = 0 \quad (y = 0),$$
$$u_0^2 \Phi_{xx} + g\Phi_y = 0 \quad (y = h_0).$$

Answer: $\Phi_n^{\pm}(x, y) = e^{\pm \alpha_n x/h_0} \cos(\alpha_n y/h_0)$ $(n = 0, 1, 2, \ldots)$, where $\alpha_n = \alpha_n(F)$ $(\pi n < \alpha_n < \pi/2 + \pi n)$ is the root of the equation $\tan \alpha = F^2 \alpha$ with the Froude number $F = u_0/\sqrt{gh_0}$.

33. The Levi–Civita variables $\tau(\varphi, \psi)$, $\theta(\varphi, \psi)$ are introduced for the velocity field $\mathbf{u} = (\varphi_x, \varphi_y)^T$ of the plane stationary potential fluid motion by the formula

$$\mathbf{u} = e^{\tau}(\cos \theta, \quad \sin \theta)^T,$$

where the potential φ and the stream function ψ are independent variables. Show that the nonlinear dynamic condition on the free boundary

$$\varphi_x^2 + \varphi_y^2 + 2F^{-2}y = \text{const} \quad (y = h(x))$$

is written in the Levi–Civita variables as

$$\theta_\psi = F^{-2}e^{-3\tau} \sin \theta \quad (\psi = \psi_0),$$

where F is the Froude number and ψ_0 is the value of the stream function on the free boundary: $\psi(x, h(x)) = \psi_0$.

34. Consider the plane potential flow of a heavy fluid inside the angle $z = re^{i\beta}$ $(r > 0, -5\pi/6 < \beta < -\pi/6)$ with a complex potential $w = \varphi + i\psi$ of the form $w(z) = Az^{3/2}$. Find a constant A for which the kinematic and dynamic conditions on the free boundary are satisfied on the angle sides:

$$\text{Im } w = 0, \quad \left|\frac{dw}{dz}\right|^2 + 2F^{-2}\text{Im } z = 0. \tag{3.43}$$

Find the modulus of the fluid velocity on the free boundary and derive the equation for the stream line L passing through the point $z = -i$ in the polar coordinates (r, β).

Remark This flow is called the *Stokes flow*.

Answer:

$$A = \frac{2}{3F}e^{i\pi/4}, \quad \left|\frac{dw}{dz}\right| = \sqrt{r}/F \quad (\beta = -\pi/6, \ \beta = -5\pi/6),$$

$$L : r^3 = \frac{2}{1 + \sin 3\beta}.$$

35. Consider the plane potential flow of a heavy fluid with free boundary having a corner point with the unknown opening angle θ_0 (without loss of generality we can assume that the corner point is located at the origin). The complex potential of the flow in a neighborhood of the corner point is looked for in the form

$$w(z) = Cz^n + o(z^n) \quad (C = \text{const} \quad n > 1).$$

Using the conditions (3.43) on the free boundary, find n and the opening angle at the corner point.
Answer: $n = 3/2$, $\theta_0 = 2\pi/3$.

36. Consider a solitary wave with profile $y = h(x)$ possessing a corner point at the wave crest with the ordinate $y = h_c$. It is known that the fluid velocity $\mathbf{u}$ vanishes at the corner point. Using the Bernoulli integral $|\mathbf{u}|^2 + 2gh = u_0^2 + 2gh_0$ (here, u_0 and h_0 are the velocity and depth of the fluid at infinity) and the value of the Froude number $F = u_0/\sqrt{gh_0} = 1.290$, found by a numerical calculation for the flow under consideration, find the dimensionless amplitude $a = (h_c - h_0)/h_0$ of the cusped solitary wave.
Answer: $a = 0.832$.

37. Show that, during the time of motion $-\infty < t < +\infty$, the solitary surface wave transfers in the direction of its propagation through each vertical cross-section of the layer the fluid mass

$$m = \rho \int_{-\infty}^{+\infty} \zeta(\xi)d\xi,$$

where $\rho = \text{const}$ is the fluid density and $y = h_0 + \zeta(x - u_0 t)$ is the free surface shape in the travelling wave (h_0 is the depth of the fluid at rest and u_0 is the wave velocity).

38. The momentum I of a travelling solitary wave type is defined by

$$I = \int_{-\infty}^{+\infty} \int_0^{h(\xi)} \rho u(\xi, y) dy\, d\xi,$$

where $u = \varphi_x(x - u_0 t, y)$ is the horizontal fluid velocity and $y = h(x - u_0 t) = h_0 + \zeta(x - u_0 t)$ is the free boundary shape. Show that $I = m u_0$, where m is the fluid mass transferred by the wave.

39. Show that the Serre–Su–Gardner system of equations

$$h_t + (uh)_x = 0,$$

$$u_t + u u_x + g h_x + \frac{1}{3h}(h^2 d_t^2 h)_x = 0,$$

$$d_t = \partial_t + u \partial_x$$

is invariant under the Galilean transformation $\widetilde{t} = t, \widetilde{x} = x - u_0 t, \widetilde{u} = u - u_0, \widetilde{h} = h$. Do the Boussinesq equations (3.18) and (3.19) possess a similar property?

40. Within the framework of the second shallow water approximation (the Serre–Su–Gardner model) find an expression for the hydrodynamic pressure $p(x, y, t)$ inside the fluid layer in terms of the functions h and u.

Hint. Integrate the vertical momentum equation and ignore terms of order higher than $O(\varepsilon^2)$.

Answer:

$$p = \rho g(h - y) + \frac{1}{2}\rho \frac{h^2 - y^2}{h} d_t^2 h.$$

41. Establish the law of conservation of energy

$$\partial_t \left(\frac{1}{2}h|\mathbf{u}|^2 + q\right) + \text{div}\left((\widetilde{p} + q)\mathbf{u}\right) = 0, \quad q = \frac{1}{2}gh^2 + \frac{1}{6}h(d_t h)^2,$$

for the Green–Naghdi system (3.20). Here, the function $\widetilde{p}$ is the same as in the law of conservation of momentum (3.21).

42. Let $h(\mathbf{x}, t)$ and $\mathbf{u}(\mathbf{x}, t)$ be solutions to the Green–Naghdi system (3.20). Trajectories of fluid particles are defined as the integral curves of the system of ordinary differential equations

$$\frac{d\mathbf{x}}{dt} = \mathbf{u}(\mathbf{x}, t).$$

Show that for an arbitrary closed contour $C(t)$ in the plane $\mathbf{x} = (x, y)$ consisting of the same particles the following relation holds:

$$\frac{d}{dt} \int_{C(t)} \mathbf{v} \cdot d\mathbf{x} = \int_{C(t)} \left(d_t \mathbf{v} + \left(\frac{\partial \mathbf{u}}{\partial \mathbf{x}} \right)^T \mathbf{v} \right) \cdot d\mathbf{x},$$

where

$$\mathbf{v} = \mathbf{u} + \frac{1}{3h} \nabla (h^2 d_t h), \quad d_t = \partial_t + \mathbf{u} \cdot \nabla.$$

43. Under the conditions of the previous problem, show that the scalar Ω/h, where Ω (a counterpart of the curl for a medium with dispersion) is defined by $\Omega \mathbf{e}_z = $ curl $\mathbf{v}$, is conserved along the particle trajectories. Here, $\mathbf{e}_z$ is the unit vector of the z-axis directed along the direction of the force of gravity.

44. Consider the linearized two-dimensional problem describing waves on the interface $y = \zeta(x, t)$ of two layers of a heavy fluid of infinite depth with the surface tension forces taken into account

$$\Phi_{1xx} + \Phi_{1yy} = 0 \quad (-\infty < y < 0),$$
$$\Phi_{2xx} + \Phi_{2yy} = 0 \quad (0 < y < +\infty)$$

with the boundary conditions at $y = 0$ (the unperturbed level of the interface)

$$\zeta_t + u_1 \zeta_x = \Phi_{1y}, \quad \zeta_t + u_2 \zeta_x = \Phi_{2y},$$
$$\rho_1(\Phi_{1t} + u_1 \Phi_{1x} + g\zeta) - \rho_2(\Phi_{2t} + u_2 \Phi_{2x} + g\zeta) = \sigma \zeta_{xx}$$

and the decay condition

$$\Phi_1(x, y, t) \to 0 \quad (y \to -\infty),$$
$$\Phi_2(x, y, t) \to 0 \quad (y \to +\infty),$$

where the constants ρ_1 and $\rho_2 < \rho_1$ are the fluid densities in the layers, u_1 and u_2 are the velocities the piecewise constant flow where the linearization is performed, and $\sigma > 0$ is the surface tension coefficient. Find the dependence of the phase velocity $c_p(k)$ on the wavenumber k. For what velocities u_1 and u_2 are all wave modes stable?

Answer: $c_p(k) = c_m \pm \sqrt{c_0^2(k) - c_*^2}$, where

$$c_m = \frac{\rho_1 u_1 + \rho_2 u_2}{\rho_1 + \rho_2}, \quad c_0^2(k) = \frac{g(\rho_1 - \rho_2)}{(\rho_1 + \rho_2)k} + \frac{\sigma k}{\rho_1 + \rho_2},$$

$$c_*^2 = \frac{\rho_1 \rho_2 (u_1 - u_2)^2}{(\rho_1 + \rho_2)^2}.$$

The stability condition:

$$(u_1 - u_2)^2 \leqslant 2\sqrt{\sigma g(\rho_1 - \rho_2)} \left(\frac{1}{\rho_1} + \frac{1}{\rho_2} \right).$$

45. Consider the problem describing waves in a two-layer fluid of finite depth with the free upper boundary:

$$\varphi_{1xx} + \varphi_{1yy} = 0, \quad 0 < y < h_1 + \zeta_1(x, t),$$

$$\varphi_{2xx} + \varphi_{2yy} = 0, \quad h_1 + \zeta_1(x, t) < y < h_1 + h_2 + \zeta_2(x, t),$$

$$\varphi_{1y} = 0, \quad y = 0,$$

$$\zeta_{1t} + \varphi_{jx}\zeta_{1x} - \varphi_{jy} = 0 \quad (j = 1, 2)$$

$$\left.\begin{array}{l} \rho_1 \left(\varphi_{1t} + \frac{1}{2}\varphi_{1x}^2 + \frac{1}{2}\varphi_{1y}^2 + g\zeta_1 \right) \\ \\ = \rho_2 \left(\varphi_{2t} + \frac{1}{2}\varphi_{2x}^2 + \frac{1}{2}\varphi_{2y}^2 + g\zeta_1 \right), \end{array}\right\} \quad y = h_1 + \zeta_1(x, t),$$

$$\left.\begin{array}{l} \zeta_{2t} + \varphi_{2x}\zeta_{2x} - \varphi_{2y} = 0, \\ \\ \varphi_{2t} + \frac{1}{2}\varphi_{2x}^2 + \frac{1}{2}\varphi_{2y}^2 + g\zeta_2 = 0, \end{array}\right\} \quad y = h_1 + h_2 + \zeta_2(x, t),$$

where $\rho_2 < \rho_1$ are the fluid densities of the layers, h_1 and h_2 are the layer depths at the equilibrium state. Linearize the equations at the state at rest $\varphi_1 = 0$, $\varphi_2 = 0$, $\zeta_1 = \zeta_2 = 0$ and derive the dispersion relation. What is the asymptotics of wave modes in the limit of close fluid densities in the layers $(\rho_1 - \rho_2)/\rho_1 \to 0$?
Answer:

$$(\omega^2 - \mu g k \tanh kh_1)(\omega^2 - gk \tanh kh_2)$$

$$+ \lambda \omega^2 \tanh kh_1 (\omega^2 \tanh kh_2 - gk) = 0,$$

$$\omega^2(k) = gk \tanh k(h_1 + h_2) + O(\mu) \text{ (modes of surface waves)},$$

$$\omega^2(k) = \frac{\mu gk}{\coth kh_1 + \coth kh_2} + O(\mu^2) \text{ (modes of internal waves)},$$

where

$$\mu = \frac{\rho_1 - \rho_2}{\rho_1}, \quad \lambda = \frac{\rho_2}{\rho_1} = 1 - \mu.$$

46. Consider linear stationary waves (i.e., wave packets with a given frequency $\omega = 0$ and the sought wavenumber k) on the piecewise constant flow of a two-layer fluid under a rigid lid with the density ρ_j ($\rho_2 < \rho_1$), velocity u_j, and depths h_1 and $h_2 = H - h_1$ in the lower ($j = 1$) and upper ($j = 2$) layers. Introduce the dimensionless parameters: the dimensionless wavenumber $\kappa = kh_1$, the ratio of unperturbed depths of the layers $r = h_2/h_1$, and the densimetric Froude numbers

$$F_1^2 = \frac{\rho_1 u_1^2}{g(\rho_1 - \rho_2)h_1}, \quad F_2^2 = \frac{\rho_2 u_2^2}{g(\rho_1 - \rho_2)h_2}.$$

Verify that, in the case of stationary waves, the dispersion relation (3.25) is written in the above dimensionless variables as

$$D(\kappa; F, r) \overset{\text{def}}{=} F_1^2 \kappa \coth \kappa + F_2^2 r\kappa \coth r\kappa - 1 = 0.$$

Show that the dispersive function $D(\kappa; F, r)$ possesses the following properties.

a. All roots of the analytic function $D(z; F, r)$ of the complex variable $z = \kappa + iv$ are located on the coordinate axes $\text{Im } z = 0$ and $\text{Re } z = 0$.
b. There are only two real roots κ with the multiplicity taken into account. These roots exist if and only if $F_1^2 + F_2^2 \leq 1$ (the case of subcritical piecewise constant flows).
c. The function $D(z; F, r)$ has countably many roots on the imaginary axis. Among these roots, there are only two roots $z_\pm = \pm i\alpha$ lying in the strip $|\text{Im } z| < \pi \min\{1, r^{-1}\}$ for $F_1^2 + F_2^2 > 1$ (the case of supercritical flows). Moreover, α is the minimal positive root of the real equation

$$F_1^2 \alpha \cot \alpha + F_2^2 r\alpha \cot r\alpha = 1$$

and $\alpha \to 0$ as $F_1^2 + F_2^2 \to 1$.

Remark The imaginary roots $\pm i\alpha$ of the dispersive function characterize the exponential asymptotics $\sim \exp(-\alpha|x|)$ as $|x| \to \infty$ for solitary waves in a two-layer fluid.

47. Using the integral laws of conservation of mass, momentum, energy and the Bernoulli integral for a piecewise potential flow, find out for what layer depths H_1, H_2, h_1, h_2 ($H_1 + H_2 = h_1 + h_2 = H$) and velocities U_1, U_2, u_1, u_2 taken as $x \to \pm\infty$ there exists a stationary two-layer flow under a rigid lid, as in Fig. 3.3, in the form of a smooth bore.

Answer:

$$\frac{h_2}{h_1} = r, \quad \frac{H_1}{h_1} = 1 + a, \quad \frac{H_2}{h_2} = \frac{r - a}{r},$$

$$\frac{U_1}{u_1} = \frac{1}{1 + a}, \quad \frac{U_2}{u_2} = \frac{r}{r - a},$$

$$F_1^2 \stackrel{\text{def}}{=} \frac{\rho_1 u_1^2}{g(\rho_1 - \rho_2)h_1} = \frac{(1 + a)^2}{1 + r}, \quad F_2^2 \stackrel{\text{def}}{=} \frac{\rho_2 u_2^2}{g(\rho_1 - \rho_2)h_2} = \frac{(r - a)^2}{r(1 + r)},$$

where $0 < r < +\infty, -1 < a < r$.

48. The equation (3.33) describing solitary waves and smooth bores in a two-layer fluid under a rigid lid has the form

$$\left(\frac{d\eta}{dx}\right)^2 = 3\eta^2 \frac{P(\eta; F, r)}{Q(\eta; F, r)}$$

with polynomials

$$Q(\eta; F, r) = F_1^2 (r - \eta) + r^3 F_2^2 (1 + \eta),$$

$$P(\eta; F, r) = F_1^2 (r - \eta) + r F_2^2 (1 + \eta) - (1 + \eta)(r - \eta),$$

where F_j ($j = 1, 2$) are the densimetric Froude numbers and $r = h_2/h_1$ is the ratio of unperturbed layer depths. Prove the following assertions.

a. The discriminant $d(F, r) = (F_1^2 - r F_2^2 + r - 1)^2 + 4r(1 - F_1^2 - F_2^2)$ of the quadratic polynomial $P(\eta; F, r)$ in η is positive if the point $F = (F_1, F_2)$ is located inside the rhombus $|F_1| + \sqrt{r}|F_2| < \sqrt{1 + r}$ and vanishes on the rhombus boundary $|F_1| + \sqrt{r}|F_2| = \sqrt{1 + r}$.

b. The roots $\eta = a_\pm(F, r)$ of the polynomial $P(\eta; F, r)$, where

$$a_\pm(F, r) = -\frac{1}{2}(r F_2^2 - F_1^2 + 1 - r \mp \sqrt{d(F, r)}),$$

satisfy the inequalities $0 < a_-(F, r) < a_+(F, r)$ if the point $F = (F_1, F_2)$ is located inside the lateral curvilinear triangles bounded by the circle $F_1^2 + F_2^2 = 1$ and the rhombus $|F_1| + \sqrt{r}|F_2| = \sqrt{1 + r}$ (cf. Fig. 3.4) and the inequalities $a_-(F, r) < a_+(F, r) < 0$ if the point $F = (F_1, F_2)$ lies in an analogous upper or lower triangle.

49. Construct a solitary wave type solution to the following approximate version of (3.33) describing nonlinear stationary waves in a two-layer fluid under a rigid lid:

$$\left(\frac{d\eta}{dx}\right)^2 = \frac{3\eta^2 \{F_1^2(r - \eta) + r F_2^2(1 + \eta) - (1 + \eta)(r - \eta)\}}{r F_1^2 + r^3 F_2^2}$$

which is obtained for waves of small amplitude if the linear denominator depending on η of the fraction on the right-hand side of (3.33) is approximated by a constant, namely, by its value at $\eta = 0$.
Answer:

$$\eta(x) = \frac{r(F_1^2 + F_2^2 - 1)}{a_\mp(F, r) \pm \sqrt{d(F, r)} \cosh {}^2 kx},$$

where

$$k^2 = \frac{3(F_1^2 + F_2^2 - 1)}{4(F_1^2 + r^2 F_2^2)}.$$

The upper sign (plus or minus) is taken for $0 < a_- < a_+$ (waves of elevation type) and the lower sign is taken for $a_- < a_+ < 0$ (waves of depression type). The quantities $a_\pm(F, r)$ and $d(F, r)$ are defined in the previous problem.

50. Consider the two-layer shallow water equations with a lid in the Boussinesq approximation:

$$A_t + ((A^2 - 1)B)_x = 0,$$
$$B_t + ((B^2 - 1)A)_x = 0.$$
$$(3.44)$$

Show that these equations are hyperbolic in the domain $|A| < 1$, $|B| < 1$. Find characteristics and Riemann invariants.
Answer:

$$\frac{dx}{dt} = 2AB \pm \sqrt{(1 - A^2)(1 - B^2)},$$

$$r_\pm = AB \pm \sqrt{(1 - A^2)(1 - B^2)}.$$

51. Under the conditions of the previous problem, show that the pair of functions $S = -(1 - A^2)(1 - B^2)$ and $F = 2ABS$ forms the conservation law

$$S(A, B)_t + F(A, B)_x = 0.$$

Find a subdomain of the hyperbolicity domain where S is convex.
Answer:

$$|B| < 1, \quad (1 - A^2)(1 - B^2) > 4A^2 B^2.$$

52. Show that all conservation laws $S(A, B)_t + F(A, B)_x = 0$ for the system (3.44) are exhausted by functions $S(A, B)$ such that $(1 - B^2)S_{BB} - (1 - A^2)S_{AA} = 0$.

53. Show that the change of variables $u = 2AB$, $h = (1 - A^2)(1 - B^2)$ transforms the system (3.44) to the shallow water equations

$$h_t + (uh)_x = 0,$$
$$u_t + uu_x + h_x = 0.$$

What is the Jacobian of the transformation?
Answer:

$$\partial(u, h)/\partial(A, B) = 4(A^2 - B^2).$$

54. Show that the system of equation (3.27) describing two-layer shallow water under a lid is hyperbolic if and only if the inequality (3.28) holds.
55. Consider the system of equations describing two-layer shallow water with a free surface

$$\mathbf{u}_t + A(\mathbf{u})\mathbf{u}_x = 0$$

for vector-valued functions $\mathbf{u} = (h_1, h_2, u_1, u_2)^T$, where $h_j > 0$ are the layer depths, u_j are the fluid velocities in the layers ($j = 1, 2$), and the matrix A has the form

$$A(\mathbf{u}) = \begin{pmatrix} u_1 & 0 & h_1 & 0 \\ 0 & u_2 & 0 & h_2 \\ g & g\lambda & u_1 & 0 \\ g & g & 0 & u_2 \end{pmatrix},$$

where $\lambda = \rho_2/\rho_1 (\rho_2 < \rho_1)$ is the ratio of densities. Find the characteristic determinant $\Delta(c) = \det(A - cI)$ of this system. Transform the characteristic equation $\Delta(c) = 0$ to an equivalent system of equations for the Froude numbers $F_1 = (u_1 - c)/\sqrt{gh_1}$, $F_2 = (u_2 - c)/\sqrt{gh_2}$. How many real roots does the equation $\Delta(c) = 0$ have if

(a) $u_2 = u_1$,
(b) $u_2 = u_1 + \sqrt{gh_1} + \sqrt{gh_2}$.

Answer:

$$\Delta(c) = ((u_1 - c)^2 - gh_1)((u_2 - c)^2 - gh_2) - g^2\lambda h_1 h_2,$$
$$(F_1^2 - 1)(F_2^2 - 1) = \lambda, \quad F_2 = \sqrt{h_1/h_2}F_1 + (u_2 - u_1)/\sqrt{gh_2},$$

(a) four roots (the system is hyperbolic),
(b) two roots (the system is not hyperbolic).

56. Construct an exact solution of the form $h_1 = h_1(t)$, $h_2 = \mathrm{const}$, $u_1 = a(t)x$, $u_2 = 0$ to the two-layer shallow water equations with a free surface (cf. the previous problem) satisfying the initial conditions $h_1(0) = h_0$, $a(0) = a_0 > 0$. Show that there exist points (x, t) in the half-plane $t > 0$ for which there are complex roots of the characteristic polynomial $\Delta(c)$ on this solution (points of nonhyperbolicity).

Answer:

$$u_1(x, t) = \frac{a_0 x}{1 + a_0 t}, \qquad h_1(t) = \frac{h_0}{1 + a_0 t}.$$

57. Show that the two-layer shallow water equations with a free surface can be represented in the form

$$\mathbf{u}_t + (R\nabla_{\mathbf{u}} E(\mathbf{u}))_x = 0$$

for $\mathbf{u} = (h_1, h_2, u_1, u_2)^T$, where R is the symmetric matrix of the form

$$R = \begin{pmatrix} 0 & 0 & \frac{1}{\rho_1} & 0 \\ 0 & 0 & 0 & \frac{1}{\rho_2} \\ \frac{1}{\rho_1} & 0 & 0 & 0 \\ 0 & \frac{1}{\rho_2} & 0 & 0 \end{pmatrix},$$

and $E(\mathbf{u})$ is the energy function

$$E(\mathbf{u}) = \frac{1}{2}\rho_1 h_1 u_1^2 + \frac{1}{2}\rho_2 h_2 u_2^2 + \frac{1}{2}\rho_1 g h_1^2 + \rho_2 g h_1 h_2 + \frac{1}{2}\rho_2 g h_2^2.$$

58. The Eulerian coordinates $\mathbf{x}$ and the Lagrangian coordinates $\boldsymbol{\xi}$ of fluid particles are connected by the relations

$$\frac{d\mathbf{x}}{dt} = \mathbf{u}(\mathbf{x}, t), \qquad \mathbf{x}|_{t=0} = \boldsymbol{\xi}.$$

Show that the system (3.1) describing the motion of an inhomogeneous fluid is written in the Lagrangian coordinates $(\boldsymbol{\xi}, t)$ as

$$\det M = 1, \qquad \rho_0 M^T (\mathbf{x}_{tt} - \mathbf{g}) + \nabla_{\boldsymbol{\xi}} p = 0,$$

where $\rho_0 = \rho_0(\boldsymbol{\xi})$ is the initial density field and $M = \partial \mathbf{x}/\partial \boldsymbol{\xi}$ is the Jacobi matrix of the mapping $\boldsymbol{\xi} \to \mathbf{x}(\boldsymbol{\xi}, t)$.

59. Show that the plane motion of a homogeneous fluid is described in the Lagrangian coordinates by the system of equations

$$x_\xi y_\eta - x_\eta y_\xi = 1,$$
$$x_\eta x_{\xi t} - x_\xi x_{\eta t} + y_\eta y_{\xi t} - y_\xi y_{\eta t} = \omega_0(\xi, \eta), \qquad (3.45)$$

where $\omega_0(\xi, \eta)$ is the initial vorticity.

60. Show that if the initial vorticity is identically constant: $\omega_0(\xi, \eta) \equiv \text{const}$, then the transformation

$$x' = x\cos\frac{\omega_0 t}{2} + y\sin\frac{\omega_0 t}{2},$$

$$y' = -x\sin\frac{\omega_0 t}{2} + y\cos\frac{\omega_0 t}{2}$$

transforms the system (3.45) to a system of the same form for $x'(\xi, \eta, t)$ and $y'(\xi, \eta, t)$ with vorticity $\omega_0 = 0$.

61. Consider the solid body rotation of an ideal incompressible fluid with constant angular velocity ω_0 about the Oz-axis with the particle trajectories

$$x = \xi\cos\omega_0 t - \eta\sin\omega_0 t,$$

$$y = \xi\sin\omega_0 t + \eta\cos\omega_0 t,$$

$$z = \zeta.$$

Find the expression in the Eulerian coordinates for the stream function ψ of the flow in the Oxy-plane ($u = \psi_y$, $v = -\psi_x$) and the vorticity vector $\omega = \text{curl } \mathbf{u}$ corresponding to the velocity field $\mathbf{u} = (u, v, 0)$ of this motion.
Answer:

$$\psi(x, y) = -\frac{1}{2}\omega_0(x^2 + y^2), \quad \omega = (0, 0, 2\omega_0).$$

62. Consider the two-dimensional motion of a homogeneous fluid in the (x, y)-plane with the particle trajectories

$$x = a + \frac{1}{k}e^{kb}\sin k(a - ct),$$

$$y = b - \frac{1}{k}e^{kb}\cos k(a - ct),$$

where $k = \text{const}$ and $c = \text{const}$, whereas the parameters a and b ($b \leqslant 0$) are constant on each fixed trajectory, but vary from one trajectory to another (the Gerstner trochoidal waves). Show that the pressure is constant along each trajectory if and only if $c^2 = g/k$; moreover, in this case, the pressure p and vorticity ω have the form

$$p = p_0 - \rho_0 gb + \frac{1}{2}\rho_0 c^2(e^{2kb} - 1), \quad \omega = \frac{2kce^{2kb}}{1 - e^{2kb}},$$

where $\rho_0 = \text{const}$ is the fluid density and $p_0 = \text{const}$.

63. Show that the law of conservation of energy

$$\partial_t\left(\frac{1}{2}\rho|\mathbf{u}|^2 + \rho gz\right) + \mathrm{div}\left(\mathbf{u}\left(\frac{1}{2}\rho|\mathbf{u}|^2 + p + \rho gz\right)\right) = 0$$

   is a consequence of the system (3.1) governing the motion of a stratified fluid.

64. At the time $t = 0$, the layer $0 < z < h_0$ of a stratified fluid has the density distribution $\rho = \rho_0(z)$ with a smooth monotone function ρ_0 ($\rho_0'(z) < 0$) and the constant pressure $p = p_0$ on the free boundary $z = h_0$. Construct a solution to the system (3.1) describing the spreading of the layer over the bottom $z = 0$ with the linear velocity field $u(x, t) = a(t)x$, $v(y, t) = 0$, $w(z, t) = -a(t)z$ ($a(0) = a_0 > 0$) under the action of the force of gravity. Find trajectories of fluid particles.

   Answer:

$$a(t) = \frac{a_0}{1 + a_0 t};$$

   trajectories of particles:

$$x = x_0(1 + a_0 t), \quad y = 0, \quad z = z_0/(1 + a_0 t); \quad \rho = \rho_0((1 + a_0 t)z),$$

$$p = p_0 - \frac{g}{1 + a_0 t}\int_{(1+a_0 t)z}^{h_0}\rho_0(s)ds - \frac{2a_0^2}{(1 + a_0 t)^4}\int_{(1+a_0 t)z}^{h_0} s\rho_0(s)ds.$$

65. Consider the atmosphere with pressure p and density ρ connected by the equation of state of an ideal gas $p = \rho RT$, where $R = \mathrm{const}$ is the gas constant and T is the absolute temperature. Using the hydrostatic law $dp/dz = -g\rho(z)$, find the dependence of the density of a gas at rest on the vertical variable z for the following temperature distributions:

   (a) $T = T_0$,
   (b) $T = T_0(1 - z/h_0)$, where $T_0 = \mathrm{const}$.

   Answer:

   (a) $\rho(z) = \rho_0 e^{-\beta z/h_0}$,
   (b) $\rho(z) = \rho_0(1 - z/h_0)^{\beta-1}$ ($\beta = g\rho_0 h_0/p_0$, $p_0 = \rho_0 RT_0$).

66. Find the density $\rho(z)$ and pressure $p(z)$ of a stratified fluid at rest in the layer $-h_0 \le z < 0$ if we know the values of the density ρ_0 and pressure p_0 on the layer surface $z = 0$ and the dependence of the Brunt–Väisälä frequency on z: $N(z) = \sqrt{\sigma g/(h_0 - \sigma z)}$ ($\sigma > 0 - \mathrm{const}$).

   Answer:

$$\rho(z) = \rho_0\left(1 - \frac{\sigma z}{h_0}\right), \quad p(z) = p_0 - g\rho_0 z\left(1 - \frac{\sigma z}{2h_0}\right).$$

67. Consider the Sturm–Liouville problem (3.36) governing linear internal waves in a layer of a fluid under a rigid lid with exponential stratification

$$\rho_0(z) = \rho_* e^{-N_0^2 z/g} \quad (\rho_* = \text{const} > 0, \quad N_0 = \text{const}).$$

Find the spectrum of frequencies and the corresponding eigenfunctions.
Answer:

$$\omega_n^2(m) = \frac{N_0^2 m^2}{m^2 + \frac{\pi^2 n^2}{h_0^2} + \frac{N_0^4}{4g^2}},$$

$$W_n(z) = e^{N_0^2 z/(2g)} \sin \frac{\pi n}{h_0} z$$

for $n = 1, 2, 3, \ldots$

68. Consider the spectral Sturm–Liouville problem

$$(\rho_0 W_z)_z = \rho_0 k^2 \left(1 - \frac{N_0^2}{\omega^2}\right) W \quad (0 < z < h_0),$$

$$W(0) = 0, \quad W'(h_0) = g \frac{k^2}{\omega^2} W(h_0)$$

describing linear internal waves in a layer of a stratified fluid with a free surface in the Boussinesq approximation. Here, $\rho_0 = \text{const}$, $N_0 = \text{const}$, and k is a wavenumber. Find the spectrum of frequencies and the corresponding eigenfunctions.
Answer:

$$\omega_0^2(k) = \frac{N_0^2 k^2}{k^2 - \lambda_0^2(k)}, \quad W_0(z) = \sinh \lambda_0(k) z,$$

where $\lambda_0(k) > 0$ is a root of the equation

$$N_0^2 \lambda \coth \lambda h_0 = g(k^2 - \lambda^2);$$

$$\omega_n^2(k) = \frac{N_0^2 k^2}{k^2 + \lambda_n^2(k)}, \quad W_n(z) = \sin \lambda_n(k) z \quad (n = 1, 2, 3, \ldots),$$

where $0 < \lambda_1(k) < \lambda_2(k) < \ldots$ are roots of the equation

$$N_0^2 \lambda \cotan \lambda h_0 = g(k^2 + \lambda^2).$$

69. Linear long waves in the shear flow of a stratified fluid under a rigid lid are described in the Boussinesq approximation by the spectral problem

$$((u_0(z) - c)^2 \varphi_z)_z + N_0^2 \varphi = 0,$$

$$\varphi(0) = \varphi(h_0) = 0,$$

where c is the phase velocity and $N_0 = \text{const}$ is the Brunt–Väisälä frequency. Find the spectrum of phase velocities and the eigenfunctions for the flow with linear shear of velocity $u_0(z) = az$ ($a = \text{const}$, $a \neq 0$) satisfying the stability condition $a^2 < 4N_0^2$.

Answer:

$$c_n = \frac{ah_0}{1 - e^{\pi n/\lambda}},$$

$$\varphi_n(z) = \frac{\sin\left\{\lambda \ln\left(1 - \frac{az}{c_n}\right)\right\}}{\sqrt{1 - \frac{az}{c_n}}},$$

$$\lambda = \sqrt{\frac{N_0^2}{a^2} - \frac{1}{4}}, \quad n = \pm 1, \pm 2, \ldots.$$

70. Show that eigenfunctions W_i and W_j of the problem (3.36) corresponding to any two real wave modes $\omega_i(m)$ and $\omega_j(m)$ with $i \neq j$ satisfy the orthogonality conditions

$$\int_0^{h_0} \rho_{0z} W_i W_j \, dz = 0,$$

$$\int_0^{h_0} \rho_0 (W_{iz} W_{jz} + m^2 W_i W_j) dz = 0.$$

71. Prove that for the spectrum of the Sturm–Liouville problem (3.36) the following estimate holds: $\omega^2 < gm$.

72. Show that the group and phase velocities of each mode of linear internal waves described by the Sturm–Liouville problem (3.36) are connected by the relation

$$c_g(k) = c_p(k) \frac{\int_0^{h_0} W_z^2(z) dz}{\int_0^{h_0} W_z^2(z) dz + k^2 \int_0^{h_0} W^2(z) dz},$$

where k is the wavenumber and W is the corresponding eigenfunction.

73. We consider the two-dimensional stationary flow of a continuously stratified fluid in the gravity field. The stream function ψ is defined by the relations

$$\psi_y = \sqrt{\rho/\rho_0}\, u,$$
$$\psi_x = -\sqrt{\rho/\rho_0}\, v,$$

where u and v are components of the velocity vector, ρ is the fluid density, and $\rho_0 = \text{const}$ is some characteristic value of the density. Show that the function ψ satisfies the Dubreil–Jacotin–Long equation in the form due to Yih

$$\psi_{xx} + \psi_{yy} + \frac{gy}{\rho_0}\frac{d\rho(\psi)}{d\psi} = \frac{1}{\rho_0}\frac{dH(\psi)}{d\psi},$$

where H is the same as in Eq. (3.38).

74. Show that the Dubreil–Jacotin–Long equation

$$\rho(\psi)\Delta\psi + \frac{d\rho(\psi)}{d\psi}\left\{ g\left(y - \frac{\psi}{u_0} \right) + \frac{1}{2}\left(|\nabla\psi|^2 - u_0^2 \right) \right\} = 0$$

is the Euler–Lagrange equation with the Lagrangian

$$L = -\frac{1}{2}\rho(\psi)|\nabla(\psi - u_0 y)|^2 + \frac{g}{u_0}\int_{u_0 y}^{\psi}(\rho(\chi) - \rho(\psi))\, d\chi.$$

References

1. Brekhovskikh, L. M., Goncharov, V. V.: Mechanics of Continua and Wave Dynamics. Springer, Berlin etc. (1985)
2. Lighthill, J.: Waves in Fluids. Cambridge University Press, Cambridge (2001)
3. Whitham, G. B.: Linear and Nonlinear Waves. Wiley, New York, NY (1999)

Supplementary References for Chapter 1

4. Godunov, S. K., Romenskii, E. I.: Elements of Continuum Mechanics and Conservation Laws. Kluwer, New York (2003)
5. Kulikovskii, A. G., Sveshnikova, E. I.: Nonlinear Waves in Elastic Media. CRC Press, Boca Raton, FL (1995)
6. Landa, P. S.: Nonlinear Oscillations and Waves in Dynamical Systems. Kluwer, Dordrecht (1996)
7. Rozhdestvenskij, B. L., Yanenko, N. N.: Systems of Quasilinear Equations and Their Applications to Gas Dynamics. Am. Math. Soc., Providence, RI (1983)
8. Serre, D.: Systems of Conservation Laws. I: Hyperbolicity, Entropies, Shock Waves. Cambridge University Press, Cambridge (1999)
9. Smoller, J.: Shock Waves and Reaction-Diffusion Equations. Springer, New York (1994)

Supplementary References for Chapter 2

10. Achenbach, J. D.: Wave Propagation in Elastic Solids. North-Holland, Amsterdam etc. (1973)
11. Karpman, V. I.: Nonlinear Waves in Dispersive Media. Elsevier (1974)
12. Nesterenko, V. F.: Dynamics of Heterogeneous Materials. Springer, New York (2001)
13. Newell, A. C.: Solitons in Mathematics and Physics. SIAM (1987)
14. Royer, D., Dieulesaint, E.: Elastic Waves in Solids I. Free and Guided Propagation. Springer, Berlin etc. (2000)
15. Sneddon, I. N.: Fourier Transforms. Courier Corporation (1995)

© Springer International Publishing AG 2017

S.L. Gavrilyuk et al., *Waves in Continuous Media*, Lecture Notes in Geosystems Mathematics and Computing, DOI 10.1007/978-3-319-49277-3

Supplementary References for Chapter 3

16. Babenko, K. I. 1987 Some remarks on the theory of surface waves of finite amplitude. Sov. Math. Dokl. 35, 599–603
17. Benjamin, T. B. 1968, Gravity currents and related phenomena. J. Fluid Mech. 31, part 2, 209–248
18. Debnath, L.: Nonlinear Water Waves. Academic Press, Boston, MA (1994)
19. Drazin, P. G.: Introduction to Hydrodynamic Stability. Cambridge University Press, Cambridge (2002)
20. Grimshaw, R. (Ed.): Enviromental Stratified Flows. Kluwer Academic Publishers, London (2001)
21. Howard, L. N. 1961 Note on a paper of John W. Miles. J. Fluid Mech. 10, 509–512
22. Johnson, R. S.: A Modern Introduction to the Mathematical Theory of Water Waves. Cambridge University Press, Cambridge (1997)
23. Kochin, N. E., Kibel', I. A., Roze, N. V.: Theoretical Hydromechanics. John Wiley & Sons, New York etc. (1964)
24. Lamb, H.: Hydrodynamics. Cambridge University Press, Cambridge (1993)
25. Milne-Thomson, L. M.: Theoretical Hydrodynamics. Mac. Millan and Co. Ltd., London (1968)
26. Miles, J. W. 1961 On the stability of heterogeneous shear fows. J. Fluid Mech. 10, 496–508
27. Novikov, S. P., Manakov, S. V., Pitaevskii, L. P., Zakharov, V. E.: Theory of Solitons. The Inverse Scattering Methods. Plenum Publ., New York (1984)
28. Pedlosky, J.: Geophysical Fluid Dynamics. Springer, New York etc. (1987)
29. Stoker, J. J.: Water Waves. The Mathematical Theory with Applications. Wiley, New York, NY (1992)
30. Sutherland, B. R.: Internal Gravity Waves. Cambridge University Press, Cambridge (2010)
31. Teshukov, V. M. 1985 On the hyperbolicity of the long wave equations. Sov. Math. Dokl. 32 469–437
32. Turner, J. S.: Buoyancy Effects in Fluids. Cambridge University Press, Cambridge etc. (1979)
33. Yih, C. S.: Stratified Flows. Academic Press, New York etc. (1980)

Index

© Springer International Publishing AG 2017
S.L. Gavrilyuk et al., *Waves in Continuous Media*, Lecture Notes in Geosystems
Mathematics and Computing, DOI 10.1007/978-3-319-49277-3

Printed in the United States
by Baker & Taylor Publisher Services